# MAYDAY!

## A HISTORY of FLIGHT through its MARTYRS, ODDBALLS AND DAREDEVILS

## DAVID DARLING

ONEWORLD

A Oneworld Book

Originally published as *The Rocket Man* by Oneworld Publications, 2013.
This paperback edition published in 2015.

ISBN 978-1-78074-409-4
eISBN 978-1-78074-565-7

Typeset by Tetragon, London
Printed and Bound in Great Britain by Clays Ltd, St Ives plc
Illustrations by Brett Ryder

Oneworld Publications
10 Bloomsbury Street, London WC1B 3SR

*With love to*
*Emily and Lewis,*
*who'll soon be flying high, too*

# CONTENTS

# LIST OF ILLUSTRATIONS

**1** 'Blanchard's Balloon' from *Wonderful Balloon Ascents* (1870) by Fulgence Marion (pseudonym of Camille Flammarion). Source: Wikipedia/public domain.

**2** An early demonstration of the Montgolfier brothers' balloon. Source: Wikipedia/public domain.

**3** Sophie Blanchard standing in the decorated basket of her balloon during her flight in Milan, Italy, in 1811 to celebrate Napoleon's 42nd birthday. Credit: US Library of Congress, Prints and Photographs division.

**4** Lincoln Beachey seated at the controls of his plane (1913). Credit: US Library of Congress.

**5** Lincoln Beachey's flight under Niagara Falls Bridge, 27 June 1911. Credit: Photo Speciality Co. (1911).

**6** Lincoln Beachey in his plane racing against Barney Oldfield, 28 June 1912. Source: US Library of Congress.

**7** Raymond Collishaw in RAF uniform (1919). Credit: RAF.

# INTRODUCTION

Do you want to jump into thin air with nothing but a pair of outsized bird wings stuck to your back, or take off clinging to a rickety framework of wood and canvas? Do you want to fly higher, faster, or further than anyone's done before in some contraption that looks like it's held together by school glue and wishful thinking? Then you're ready to join the club of pioneering aviators: that band of daredevil adventurers who have risked life and limb to push back the boundaries of flight.

But don't expect to live long. The survival rates for those who went up in the early balloons and planes aren't encouraging. Many of the characters in this book ended up in fatal or near-fatal crashes – taking one risk too many. Hydrogen balloons caught on fire and plummeted to the earth, early winged craft flipped or broke apart or just plain fell out of the sky, pushed beyond their limits. The German Otto Lilienthal – the 'Glider King' – lasted longer than most. He built and tested a variety of his own craft at the end of the nineteenth century and even made an artificial hill near Berlin as a launch pad. Between 1891 and 1896, he and his brother Gustav flew about 2,000 times, risking death every time they leapt off the slope. Eventually, Otto's luck ran out: his glider stalled, and he fell more than fifty feet, snapping his spine. He died the next day uttering the final words: '*Kleine Opfer müssen gebracht werden*' ('Small sacrifices must be made').

The fatality rate among early exhibition aviators was terrifyingly high – around ninety per cent. They not only vied to out-do each other, they were flying planes at a time when designers were still battling to understand the most basic problems in aeronautics. Arch Hoxsey and Ralph Johnstone starred in the Wright brothers' exhibition team formed in the spring of 1910. For a brief time they enthralled crowds around the States and were dubbed the 'Heavenly Twins' by newspapers, but by the end of the year both had been killed: Johnstone when his aircraft's wing's broke off during a 'spiral glide' and Hoxley while trying to set a new altitude record. Working for the rival team of Glenn Curtiss, Charles 'Daredevil' Hamilton flew dirigibles and made death-defying parachute jumps. No stunt was too outrageous for him and, incredibly, he survived more than sixty crashes, though he was permanently scarred, had two replacement silver ribs, and needed metal plates in his skull and shin.

Another member of the Curtiss troupe, Lincoln Beachey, is one of the heroes of this book. Famed as the 'The Man Who Owns the Sky', he was aviation's biggest money-spinner – a rock star of the air who couldn't get enough adulation from the crowds that gathered wherever he performed. He was the first to fly upside-down and the first American to do a loop-the-loop. His 'Dip of Death' involved diving full tilt at the ground before pulling up at the last moment. In a single year, between 1913 and 1914, around seventeen million people watched his jaw-dropping stunts in more than 120 cities. But, in March 1915, fate caught up with him when a new monoplane he was flying broke up and smashed into the waters of San Francisco Bay.

At the start of the First World War, biplanes driven by piston engines were used to carry out scouting missions over enemy territory. A couple of years later, they'd become manoeuvrable enough for hair-raising air-to-air combat and the age of the flying ace had arrived. This was the time when the Red Baron, Manfred von Richthofen, and other skilled fighter pilots, friend and foe, became household names.

By the end of the First World War, the commercial potential of the aeroplane was blindingly obvious to everyone involved in flight, and the period between the world wars is often referred to as the Golden Age of Aviation. This was the era of the barn-stormer, the wing walker, and the great air races in which speed records were smashed year after year. The skill-cum-madness extended to dancing, target shooting, and playing tennis on the wings, hundreds of feet up, while in the background the civil aviation industry began to flex its muscles.

As time went on, planes flew not only further, but faster, and – especially during dogfights – in extraordinary high-speed manoeuvres. Pilots were subjected to more and more g-forces or 'gees'. Even before 1920, aviators knew about the menace of G-LOC – g-induced loss of consciousness – in which the plane's acceleration, in a tight turn for instance, could cause blood to drain from the head and induce a brief but potentially fatal faint. G-LOC came to the fore with the development of fast monoplanes just before and during the Second World War, and following the arrival of the jet. To come to grips with its effects and find a means to counter it, subjects were whirled around in centrifuges and put through all kinds of other stomach-churning tests.

Powerful jets and rocket planes took humans past the speed of sound, then Mach 2 and Mach 3. Test pilots flew to the edge of space in vehicles whose wings were built for extraordinary speed, but not stability. Some of these pilots also ballooned into near airless blackness, tens of miles above the ground, and then leapt out of their cramped metal gondolas, with the very curvature of the Earth in view, plunging through far sub-zero temperatures, until finally they opened their parachutes as they crossed into the denser regions of the atmosphere to break their fall.

Today there are new heroes and heroines of the air: balloonists and pilots of ultralight planes, who circumnavigate the globe in journeys lasting days or weeks; astronauts, who not only fly faster than anyone through the atmosphere, but also hurtle far beyond it to orbit the planet or land on other worlds. And still there are the eccentrics, the one-of-a-kinds, who are willing to strap a rocket-pack to their back and fly with nothing else other than wings attached to their arms, like the birdmen of old.

# 1

# THE ODDEST COUPLE IN THE AIR

'Test pilot wanted. Candidates should be timid, shy, physically frail, with no previous flying experience.' Not the most likely job ad you'll ever come across. But the chances of Marie Madeleine-Sophie Armant ever becoming a pioneering aviator must have seemed about as remote. That is until, in 1804, she became the second wife of Jean-Pierre Blanchard, twenty-five years her senior, celebrated early balloonist, and all-round disagreeable character.

A great aerial pioneer and stuntman he may have been, but Jean-Pierre was also egotistical, mercenary, and not averse to stabbing a colleague in the back if it helped further his own career. Having left his parents' rural home as a young teen, tired of the poverty he'd grown up with, he wound up in Paris as a mechanic and part-time inventor. While still a boy, he devised a rat trap that involved a pistol, a hydraulic pump that could lift water 400 feet out of the River Seine, and an early form of bicycle called a velocipede. A few years later, he became obsessed with flight. If birds could manage it, thought Blanchard, why not humans? So he came up with a *vaisseau volant* ('flying vessel') that used foot pedals and hand levers to flap four bird-like wings. It had about as much chance of getting off the ground as a hippo with a propeller, but that didn't stop Blanchard from claiming to have flown it when no one was around. His opportunity to become a true and celebrated aeronaut, though, was soon to come.

## GAS BAGS AND INFLATED EGOS

On 21 November 1783, Jean-François Pilâtre de Rozier and François Laurent, the Marquis d'Arlandes, made the dramatic first flight in a hot-air balloon, built by the Montgolfier brothers. The era of human aviation had arrived. France was, for a while at least, the ballooning capital of the world, and Jean-Pierre Blanchard saw his big chance for fame and fortune. His own balloon would be lifted skyward not by hot air but by hydrogen – the lightest of gases and the most lethal if it caught fire. To the gondola he attached wings powered by oars, in the futile hope that they would help him steer. On 2 March 1784, the thirty-one-year-old Frenchman, accompanied by a monk name Pesch, climbed into his weird vehicle, moored in the Champ de Mars park in Paris. Brother Pesch was something of a rebel, having just escaped imprisonment at the hands of his order and become a would-be aeronaut in defiance of a command forbidding his travel in this 'invention of the Devil', designed – his closeted companions argued – to undermine belief in miracles.

A big crowd had gathered to watch the launch. But just as final preparations were under way, a young man dressed as an officer in the French Army forced his way to the front. He was Dupont de Chambon, a chum of Napoleon Bonaparte from his military training days, whose request to go with Blanchard on this much-publicized jaunt had already been spurned. Now he violently demanded to be let aboard and, when Blanchard refused, began hacking at the mooring ropes and steering flaps with his sword. Before police could drag him away he also managed to stab Blanchard in the hand. After things had calmed down, Blanchard, evidently put off the idea of taking passengers, decided to fly solo, much to the dismay of Pesch

**1** 'Blanchard's Balloon' from *Wonderful Balloon Ascents* (1870) by Fulgence Marion (pseudonym of Camille Flammarion).

who, for his troubles, was banished to his order's most remote monastery.

Once airborne, Blanchard, ever the optimist, tried manfully to 'row' north-east towards the commune of La Villette. But the laws of aerodynamics – of which he had a feeble grasp – and a contrary blowing wind forced him across the Seine to Billancourt, where he landed unceremoniously in the Rue de Sèvres. The Parisian press, having witnessed the bombastic aeronaut's manic attempts to row in exactly the opposite direction to which he was compelled to go, had fun playing with Blanchard's adopted motto – *Sic itur ad astra* ('Thus you shall go to the stars').

But Blanchard was no idiot. France was in the grip of balloon mania. There were pictures of balloons on everything from ceramics to fans and hats. Hair was styled *à la montgolfier* or *au demi-ballon*. Any lady who was anyone wore clothing *au ballon* with outrageously billowed skirts and puffed sleeves. Every inventor and daredevil in the land wanted to take to the air. Blanchard realized that to make a name for himself, and a fortune to boot, he'd have to try his luck abroad, where there was less competition, and so in August 1784 he moved to England.

Thanks to an over-the-top publicity campaign in which he claimed the mantle of 'world's greatest aeronaut', he won the backing of a number of wealthy patrons. One of these was John Jeffries, a Bostonian physician living in England, who agreed to finance Blanchard's attempt at the first aerial crossing of the English Channel – providing that he could come along for the ride. Although the Frenchman was eager for his benefactor's cash, he certainly didn't want to have to share the glory with him. However, Jeffries insisted, even adding a clause into the contract to the effect that if his extra weight jeopardized the success of the flight, he would consider himself expendable and bail out in mid-air.

## TWO MEN IN A BALLOON

With his sponsor's money safely banked, Blanchard did everything he could to avoid Jeffries making the trip. At the launch site of Dover Castle on England's south coast, Blanchard set up a barricaded camp that Jeffries was forced to storm with the help of some hired sailors. The two men appeared to reconcile after this fracas but the wily Frenchman had one last trick up his sleeve – or, rather, around his waist. On 7 January 1785, with the balloon inflated and ready to go, Blanchard declared that it was overweight and couldn't possibly ascend with Jeffries aboard. Suspecting foul play, the good doctor insisted on searching the slightly built hero of the air and found him to be wearing a lead belt. Finally, unencumbered by surplus metal, the balloon took off with its uneasy crew of two, heading south-east under a gentle breeze.

Inevitably, it wasn't long before Blanchard and Jeffries fell out – perhaps almost literally, given the violence of the dispute, which centred on the relative merits of their nations of birth. Both were fiercely patriotic. Hostile words and insults were exchanged and the long and short of it was that both their nation's flags, having been proudly displayed for all the world to see at lift-off, ended up in the sea with their respective owners fuming at the loss.

Eight miles out over the Channel the pair found themselves descending prematurely. Hurriedly, they tossed some ballast overboard but still the balloon headed down. Another argument broke out over what next to sacrifice. Possibly the terms of the contract were discussed. But as the cold waters came ever closer, the two men, neither of whom could swim, continued to eject only inanimate cargo. With the French coast now in sight but the gondola just a few feet above the sea, the ropes, anchors, seats, and scientific instruments were jettisoned for the sake of elevation. Blanchard even stripped to his underwear and tossed his clothes overboard. At first Jeffries baulked at disrobing, saying he would rather face a watery grave than the French *dishabille*. But as the frigid waters beckoned, Jeffries not only peeled down to his long johns but, having clambered like a rat into the rigging, offered his professional opinion that they should both empty their bladders and perhaps more to relieve the situation.

All the desperate weight shedding worked – but too well. As the balloon neared the shores of France, a warm updraft swept the scantily dressed aeronauts skyward, and without any landing ropes or anchors by which to catch hold of terra firma they climbed high again. For eleven miles they drifted inland until, at last, over a forest near the town of Guînes, Jeffries was able to grab hold of some passing tree branches. As the balloon slowed and drifted over a clearing, some of its hydrogen was released and the adventure came to a tame conclusion. Reclothed by well-wishers on the ground, the intrepid pair were borne by carriage to Calais where they were greeted by cheering crowds. Although one of the items ejected had been the mail bag, Jeffries had managed to stuff a single letter, addressed to Temple Franklin, Benjamin Franklin's grandson, into his underwear. By such means was the first airmail letter delivered.

**2** An early demonstration of the Montgolfier brothers' balloon.

# BALLOONING – FOR THE RECORD

**1783**

19 Sep    First balloon to carry passengers – a sheep, a duck, and a hen – demonstrated by the Montgolfier brothers for King Louis XVI.

21 Nov    First recorded manned flight in a hot-air balloon built by the Montgolfier brothers.

1 Dec     First hydrogen-balloon flight by Professor Jacques Charles and the Robert brothers.

**1785**

7 Jan     First balloon crossing of the English Channel by Jean-Pierre Blanchard and John Jeffries.

**1793**

9 Jan     First manned balloon flight in North America by Jean-Pierre Blanchard.

**1852**

24 Sep    Flight of the first steerable balloon, or dirigible, by Henri Giffard.

**1931**

27 May    Auguste Picard and Paul Kipfer became the first to reach the stratosphere in a balloon.

**1933**

31 Aug    Alexander Dahl took the first picture of the Earth's curvature in an open hydrogen balloon.

**1978**

16 Aug    The *Double Eagle II*, and its three-man crew, became the first balloon to cross the Atlantic.

**2002**

25 May    Altitude record set for an unmanned balloon – 53 kilometres (173,882 feet) – launched by the Japanese space agency, JAXA.

**2012**

14 Oct    Current altitude record for a manned balloon set at 38,960.5 metres (127,823 feet) by Felix Baumgartner in the Red Bull Stratos balloon.

## UPS AND DOWNS

On his return to England, Blanchard, now a celebrity, went into the business of putting on balloon shows and stunts. The parachute had recently been invented, and Blanchard entertained his audiences with displays of animals descending gently from his balloons – gently, that is, for the most part. Sadly, this particular spectacle lost its appeal after a dog and a sheep plunged to their doom. Unfazed, the ingenious but ethically challenged Frenchman sold tickets at a fancy price to watch a violinist play his instrument during a parachute jump. But when the musician leapt out less than ten feet above the ground, and managed only a handful of frantic strokes of his bow while in the air, the crowd grew ugly.

Seeing that the time was ripe to move on, Blanchard headed for Europe where he recorded the first balloon flights in Belgium, Germany, the Netherlands, and Poland. In 1788, he wowed onlookers in Basel, Switzerland, by cutting free the basket under his balloon in order to gain height and then hanging on from the dangling ropes for the rest of the trip. Five years later, in Philadelphia, Blanchard gave North America its first taste of lighter-than-air human flight in front of an audience that included President George Washington and future presidents Adams, Jefferson, Madison, and Monroe.

By this time, Blanchard had already abandoned his first wife, Victoire Lebrun, and their four children in favour of his total immersion with international ballooning exploits, leaving Victoire to die later in poverty. In 1804, he married twenty-six-year-old Sophie Armant, a person so nervous that she startled at loud noises and was afraid to ride in horse-drawn carriages. Heights, by contrast, didn't seem to bother her, and the quiet calm of floating in the air may even have come as a welcome relief. At any rate, she accompanied her husband on his aerial jaunts right from the start, joining him for the first-ever honeymoon trip in the sky.

Jean-Pierre, for all his ambition and imaginative schemes, was a lousy businessman and he looked on his young spouse as a way of pulling in more fans and much-needed cash. Seeing a woman in a balloon was still a novelty, and though Sophie may not have been the first of her sex to fly she was the first to become a professional balloonist and the first to fly solo. On only her third ascent, on 18 August 1805, she took off alone from a garden next to the cloister of the Jacobin Church in Toulouse – a useful preparation, as it turned out, for what was to come.

In February 1808, during his sixtieth balloon flight, Jean-Pierre suffered a heart attack while airborne over The Hague. He tumbled out of his basket and fell fifty feet, suffering injuries so severe that he never recovered from them. He died just over a year later.

## SOPHIE GOES SOLO

To support herself and help pay off the debts left by her profligate partner, Sophie launched into a solo aeronautical career. Hydrogen balloons were her conveyance of choice because, although more dangerous than the hot-air variety, they were easier to handle. She didn't need to tend a fire to stay airborne and, with her slight build and a basket no bigger than a chair, could use the buoyant hydrogen design to rise easily in balloons of even modest size.

**M. S. BLANCHARD** *celebre aeronauta*

*al momento del volo aerostatico da Lei eseguito in Milano*

*in presenza delle L.L. A.A. I.I. e R.R.*

*la sera del 15. Agosto 1811.*

**3** Sophie Blanchard standing in the decorated basket of her balloon during her flight in Milan, Italy, in 1811, to celebrate Napoleon's 42nd birthday.

Pretty soon, Sophie was the toast of Europe and, everywhere she went, large crowds came out to watch. Napoleon made her 'Aeronaut of Official Festivals', which meant she was in charge of organizing balloon displays at all the major ceremonies in France. In 1810, she flew over the Champs de Mars (near where the Eiffel Tower is today) to honour Napoleon's marriage to Marie-Louise of Austria. To commemorate the birth of their son, she again flew over Paris, dropping announcements of the event. A year later, during official celebrations of the boy's baptism, she ascended above the Château de Saint-Cloud, a magnificent palace overlooking the Seine, west of the French capital, and entertained spectators with what would become her signature trick – setting off fireworks from her balloon hundreds of feet above the ground.

Evening flights were Sophie's speciality. The air was calmer then and, as the sky darkened, her pyrotechnics could be seen to their best advantage. But it was a horrendously risky venture, working with flames so close to a big bag of the most explosive gas on the planet. The fireworks were contained in small baskets and lighted on a fuse before being allowed to drift down by parachute.

Going up in smoke was only one of the dangers that Sophie faced. She didn't go in for quiet evening jaunts within shouting distance of the ground, she flew at heights of more than 10,000 feet, endured sub-zero temperatures, and sometimes blacked out from the altitude and cold. On one occasion she had to stay high in the air for over fourteen hours to avoid a hailstorm that was going on below. On another, she narrowly avoided drowning when she crashed into a marsh.

But, despite the ever-present dangers of her job, she outlived Napoleon's time in office. Not only that but she seamlessly shifted allegiance and became a favourite of the returning royalty in the person of Louis XVIII. In 1814, she was on hand to celebrate Louis' return to the throne, ascending in style from the Pont Neuf. So impressed was the king by her performance that he gave her the slightly amended title of 'Official Aeronaut of the Restoration'.

Inevitably, Sophie's luck finally ran out. It happened on the evening of 6 July 1819, on her fifty-ninth flight – just one short of her husband's career tally. Everything started out normally, although for some reason Sophie seemed ill at ease. She had been warned plenty of times in the past about setting off fireworks near her balloon. Perhaps it was the stiff breeze that was causing her some concern. At any rate, she was determined to go ahead with the display and, as usual, was dressed to the nines for the occasion: a long white dress and white hat topped with ostrich feathers.

Up she went, waving a white flag at the enthralled onlookers in the Jardin de Tivoli. But from the outset the wind proved to be a problem, driving the balloon sideways into a tree. To gain height more quickly, Sophie threw out ballast, at the cost of some stability. Finally she cleared the obstructions on the ground, rose high into the air, and began her pyrotechnic show by setting off some Bengal Fire – a mixture of substances that burn with an intensely bright flame – to illuminate her balloon. But something went wrong. A spark from the fire reached the hydrogen, catching it alight. The crowd below, not realizing what was happening, thought for a while that the brilliant spectacle was part of the performance and burst into applause. Meanwhile, Sophie was frantically tossing out more ballast to keep herself airborne. It was a losing battle: the balloon rapidly lost buoyancy while, at the same time, the wind carried it more and more off course. In horror, the crowd watched as the balloon drifted above the buildings of the Rue de Provence, until, in the final moments, the hydrogen completely burned up and the charred envelope dropped onto a high rooftop. Even then Sophie might have survived, but a sudden gust caught the deflated cloth and tipped the aeronaut out of her small basket and to her death in the street below.

The crowd was stunned, and the owners of the Jardin de Tivoli immediately donated the admission fees to the support of Blanchard's children. When they found out that she didn't have any offspring, the money was used to build a memorial to Sophie over her grave, on which was engraved the epitaph *victime de son art et de son intrépidité* ('victim of her art and intrepidity').

# 2

# Insanity in a Pinstripe

Just eight years after the Wright brothers' historic first flight in 1903, a sharply dressed pilot loaded up the gas tanks of his flimsy-looking plane and announced to the watching crowd that he intended to fly straight up into the sky till his fuel ran out. For over an hour and three-quarters he spiralled upwards until finally his engine sputtered and died. Then, in what seemed

**4**   Lincoln Beachey seated at the controls of his plane (1913).

15

like a suicidal plunge, he tumbled back down before levelling out just above the ground and gliding to a safe landing. The plane's pressure gauge showed that he had peaked at 11,578 feet – a new world altitude record. That's how Lincoln Beachey rolled. One of the most naturally gifted aviators who ever lived, he was arrogant, brilliant, and a consummate showman.

Born in San Francisco in 1887, Beachey began tinkering with machines at an early age and opened his own bicycle shop when he was thirteen. A couple of years later he graduated to fixing motorcycles and their engines. Soon after that, he joined the performing stunt team run by pioneering balloonist and former circus trapeze artist Thomas Baldwin, starting out as a mechanic and quickly moving on to becoming a pilot as well. In 1900 Baldwin got into the business of building dirigibles – rigid airships powered by small engines. Beachey helped Baldwin put together a dirigible called *California Arrow*, propelled by a motorcycle engine supplied by Glenn Curtiss, who would go on to found the US aircraft industry. In 1905, still only seventeen, Beachey took the controls of the *California Arrow* and made his first solo flight. There and then he was hooked, and shortly after went into the dirigible business himself, advertising his airship in a style that no one would get away with today – flying it around the Washington Monument and down the Mall, before cheekily touching down on the White House lawn.

More and more obviously, though, the future lay with planes and, grasping this, Beachey signed up for lessons at the Curtiss Flying School in San Diego. He quickly mastered the Curtiss Model D biplane – the first aircraft in the world to be built in any quantity – and was put on the company's exhibition team. But like many geniuses, Beachey was as stubborn as he was talented and managed to smash up several planes while attempting extreme manoeuvres before emerging as the team's star performer and Curtiss's top money-spinner. Whenever he flew he dressed as if he was on a big night out, complete with pinstripe suit, high collar, fancy tie, and golf cap turned fashionably backwards. Crowds would gasp as he did his signature stunt – the 'Dip of Death'. In this, he would climb his plane to 4,500 feet, then plummet towards the ground at full speed with his arms outstretched. At the last moment he levelled out, gripping the control stick with his knees and waving with both hands to his adoring fans.

The years 1908 to 1915 spanned the 'exhibition era' of early aviation, when promoters staged extraordinary stunt shows to enthuse the public in the new heavier-than-air flying machines. This was the time when people started to get excited about not just the tricks that aircraft could perform but the possibilities for their practical uses in the future. It was the age when the focus of attention shifted from airships to aeroplanes. And the greatest of the exhibitioners – the man who commanded the biggest salary and audiences – was Lincoln Beachey.

## NO STUNT TOO EXTREME

Beachey first won international fame when, on 27 June 1911, he diced with death above Niagara Falls. Taking off from an airport in New York, he arrived at the falls and circled several times over an estimated 150,000 spectators before plunging down towards the mist and spray in the gorge. Pulling up just twenty feet above the turbulent river, he then continued down the gorge before astonishing onlookers by flying under the arch of Honeymoon Bridge (also known as Fallsview

**5** Lincoln Beachey's flight under Niagara Falls Bridge, 27 June 1911.

Bridge). Later that year, he flew above a speeding train and brushed his wheels against the top of the carriages as they passed beneath him.

With every stunt, Beachey grew more bold and ambitious. In a profession with a devastatingly high mortality rate, he wanted to push himself and his machine to the absolute limit. In 1913, he became the first to fly a plane entirely indoors. Taking off inside the Machinery Palace on the Exposition grounds at the San Francisco World's Fair, he circled his plane around and around at sixty miles per hour before landing again, all within the confines of the hall.

Whenever Beachey heard of others who had pioneered new stunts, he immediately wanted to master the tricks himself and then give them his own special twist (sometimes quite literally). On 21 September 1913, a French aviator, Adolphe Pégoud, working as a test pilot for the great Louis Blériot, performed an inside loop – flying horizontally at top speed before pulling back on the control stick and doing a complete vertical circle. Pégoud thought his loop was a world's first but it turned out that he'd been beaten to it, twelve days earlier, by a Russian military pilot called Pyotr Nesterov at an army airfield near Kiev. Both men had flown monoplanes (single-wing aircraft) that were more manoeuvrable and equipped with more powerful engines than anything that Beachey had access to.

The American was desperate to do the loop himself and urged Glenn Curtiss to build him a plane that would make it possible. When Curtiss refused, Beachey stormed out of the flying team and, in what may have been more a fit of pique than anything else, composed a melodramatic essay in which he wrote: 'I was never egotistical enough to think that the crowds came to witness my skill in putting a biplane through all the trick-dog stunts. There was only one thing that drew them to my exhibitions … They paid to see me die. They bet I would, and the odds were always against my life, and I got big money for it.' On 7 March 1913, he announced that he had quit professional flying because so many young aviators had been killed trying to copy his stunts.

## INS AND OUTS AND LOOP-THE-LOOPS

Beachey's retirement proved to be brief, however. Within a few months Curtiss relented and stumped up the funds to build a plane powerful enough to tackle the loop. On 7 October, Beachey was behind its controls at Hammondsport, New York, ready to add to his aerobatic prowess. But the flight went badly wrong. As Beachey banked around over the airfield, shortly after take-off, a downdraft or possibly a mechanical glitch caused the plane to dip, sweeping two naval officers and their lady companions off the roof of a hangar from which they had been hoping to get a perfect view of the event. Beachey's plane ended up a mangled wreck in a nearby field, although the aviator himself escaped with only minor injuries. Not so fortunate was one of the women, who died from her fall.

Beachey quit aviation a second time after the accident but, again, only temporarily. A circus poster depicting a plane flying upside down – a stunt never before attempted – fired his imagination. He simply had to get back into the cockpit and master both the loop and flying horizontally with his head pointing at the ground. This time he opted to go it alone, without the backing of a big company. After conquering the loop, which he went on to repeat many hundreds of times, he wrote in his usual theatrical style: 'The silent reaper of souls and I shook hands that day. Thousands of times we've engaged in a race among the clouds. Plunging headlong in to breathless flight, diving and circling with awful speed through ethereal space … I have imagined Him close at my heels. On such occasions I have defied him … Today, the old fellow and I are pals.'

Beachey's problem was that he needed to make a living by charging people to come to his displays, so he did his flying only over exhibition grounds to which there was an admission fee. But because the stunts took place well off the ground, people could just as easily watch for free from a distance – which is exactly what many of them did. Frustrated by his lack of income, the imperious airman retired for a third time. But, unsurprisingly, not for long.

## RACE TO FAME

It so happened that Beachey shared a publicity agent with a well-known racing-car driver called Barney Oldfield. The agent hatched an ingenious plan that would put both Beachey and Oldfield firmly in the public eye *and* force people to pay if they wanted to see the action. Eye-catching posters were printed in which the 'Demon of the Sky' was shown pitted against the 'Daredevil of the Ground'. All over the country, on courses surrounded by a high fence to ensure there were no

**6** Lincoln Beachey in his plane racing against Barney Oldfield, 28 June 1912.

21

freebies, Beachey and Oldfield raced for 'The Championship of the Universe' – Beachey flying a new plane he had designed and built himself, called the 'Little Looper', which had B-E-A-C-H-E-Y painted in three-foot-high letters on its upper wing, against Oldfield's famous red 100-horsepower Fiat. Crowds flocked to see the spectacle – 30,000 strong in Dayton, Ohio, home of the Wright brothers – and thrilled as the two vehicles, throttles open wide, unmuffled engines roaring, sped past the grandstand, with clouds of dust billowing from the wheels of the big car. Although Beachey's plane was faster, the whole affair was staged so that the stars of the show knew exactly who would 'win' each time. After the pair crossed the finishing line, always nearly neck and neck, Beachey would soar up into the sky and put on a breathtaking aerobatic display – adding more and more loops as time went on, in order to keep ahead of the records set by other stunt pilots. Eventually, he was doing as many as eighty loops, one right after the other.

As well as his mock races with Oldfield, Beachey took his solo aerial shows to no fewer than 126 cities around the US between May and December 1914. In total, 17 million Americans turned out in that frenetic seven-month period to watch the Little Looper put through its paces, and everywhere he went the 'Alexander of the Air', as the press hailed him, was given superstar treatment. In the summer of 1914, Beachey also achieved another of his long-held ambitions – to fly in front of the great Thomas Alva Edison.

'I want to show such men as Henry Ford, Thomas Edison and other inventive and manufacturing geniuses', he said, 'how I handle the Little Looper. I do not believe they dream such things are possible … I want to open the eyes of the people to the possibilities of the aeroplane. My tour this summer will help advance the science of flying by ten years.'

He certainly made a lasting impression on Edison. 'I was startled and amazed,' said the inventor, 'when I saw that youngster take to the sky and send his aeroplane through the loop and then follow that feat with an upside-down flight. I could not believe my own eyes, and my nerves were a tingle for many minutes.'

Another famous engineer who came in for a surprise that summer was Orville Wright. He'd earlier doubted that Beachey actually flew loop-the-loops or upside down. 'It is probably done high in the air,' said Wright, '3,000 feet or so, an optical illusion and promoter's hype.' Some of his scepticism probably stemmed from a feud that had been going on between Glenn Curtiss and the Wright brothers for several years and that had come to a head after Wilbur Wright's death in 1912 – an event that Orville attributed to stress from the dispute. But all doubt about what Beachey could do evaporated when he saw the stuntman fly first-hand and watched the incredible finale to his show. Loop after loop Beachey performed, then flipped his craft over, flew it upwards until it stalled, fell backwards, tail first and upside down, brought his tail up until he stalled backwards, upside down, and repeated the whole manoeuvre, again and again, falling out of the sky as he tipped back and forth – all with his arms stretched wide, operating the plane with his knees and body alone. Wright couldn't believe his eyes: 'An aeroplane in the hands of Lincoln Beachey is poetry … His performance not only surprised me, but amazed me as well. He is more magnificent than I imagined.'

## INVASION WASHINGTON

Beachey did a huge amount to promote aviation in the US, but more than anything he was desperate to persuade the powers that be to invest more in military planes. In 1914, the Russian armed services had over 1,500 aircraft at their disposal, France and Germany each had over 1,000, and even Mexico had 400, while the US could only muster a mere 23. Beachey distributed millions of brochures around the country, urging people to put pressure on their representatives to argue for more military investment in aerial forces. He also invited government officials to a demonstration of how effective a plane could be in combat. But when only two members of the cabinet showed up for the event, he decided to take matters into his own hands – and buzz the White House and Congress in a mock attack.

According to one account, President Woodrow Wilson was at work in the Oval Office when he heard a noise that at first he took to be a fly. As the buzzing grew louder, he realized it was coming from outside and looked out through the window to see a biplane heading straight towards him. The machine pulled up at the last moment and continued with repeated 'attacks' on the White House, Mall, and Capitol.

Some of the details of Beachey's fake assault on the heart of Washington may be apocryphal. But there's no doubt that the event took place and that he created a stir that day with his demonstration of how deadly the plane could be as a weapon of war. As news of his antics quickly spread, crowds spilled out onto the streets, including many lawmakers who had adjourned from Congress to see what was going on. After the flight, Beachey hammered home his message: 'If I had had a bomb, you would be dead. You were defenceless. It is time to put a force in the air.'

The demonstration had its effect and Congress voted shortly after to boost spending to build up a fledgling air force. Beachey was lauded for giving his dramatic wake-up call and was offered a top post in the government's new aviation set-up. But although he turned this down because of other commitments, he carried on with his military propaganda shows.

In 1915, ahead of the Panama-Pacific International Exposition, Beachey had a large wooden model made of the USS *Oregon*, an elderly battleship, and anchored it one mile out in San Francisco Bay. The US Navy was in on the stunt and agreed to let 100 sailors man the fake vessel, which was loaded with explosives. A crowd of 80,000 on the shore, not realizing that the ship was just a mock-up, gazed in horror as Beachey flew over it in his new, bird-like Taube monoplane and dropped what appeared to be a live bomb, trailing smoke as it sped towards its target. Explosions ripped through the ship – the crew having already secretly departed aboard a tugboat – and many onlookers thought for a while that they were witnessing a horrible accident, or possibly an act of mass murder.

## GRAND FINALE

A few days later, on 14 March 1915, Beachey took off on what would be his last flight. With the Expo now in full swing, a quarter of a million people thronged the fairgrounds and sur-rounding hills. They watched as Beachey rolled his powerful new plane onto its back and began a breathtaking inverted run barely 2,000 feet above the waters of San Francisco Bay.

Perhaps he was flying lower than he thought. At any rate, it was clear that the aircraft was losing height and, in an attempt to flip through 180 degrees and gain some altitude, Beachey pulled hard on the controls. So great was the strain on the wings that both of them sheared clean off and the fuselage – with Beachey still aboard – plunged into the Bay and sank. An agonizing hour and three-quarters later, US naval divers were able to drag the airman from his submerged cockpit and, even though it was hopeless, attempts were made for several hours to try and revive him. An autopsy revealed that he had survived the impact but died from drowning.

News of Beachey's death raced around the world and San Francisco's phone system was jammed for twenty-four hours. The city's mayor took personal charge of arrangements for the funeral and tens of thousands turned out to pay their respects to one of the greatest names in aerobatics. Preparations were made for a string of grand memorials and ceremonies, which ought to have helped immortalize Lincoln Beachey's name. Yet few have even heard of him today. The simple fact is that his life and death were largely forgotten because of a string of other events that gripped the public's attention. He died just at the start of the First World War and most plans to honour him were delayed by years, by which time the memories of his exploits had faded. New heroes emerged out of daredevil exploits during the European conflict. Then there followed the record-breaking exploits of the likes of Charles Lindbergh, and yet another major public concern – the Great Depression – which helped push the extraordinary adventures of Beachey into obscurity.

Beachey was unique to his time, just as the exhibition era was unique to its time and place. The barnstorming of the

1920s and the great air races and air shows of the 1930s were different in many respects from the great aviation meets and exhibitions of the pre-First World War period. The exhibition era was a time of much experimentation and purposeful risk, and it has become a time almost lost to memory.

# 3

# BLACK ACE

Few things sharpen your reactions and ingenuity like being involved in aerial shoot-outs, day after day. Even if it takes some extreme manoeuvre to get out of the line of fire, or turn the tables on your opponent, you're going to try it, despite the risk, rather than end up in a heap of burning wreckage on the ground.

Dogfighting started during the First World War as planes became faster and nimbler and began to carry weapons. At the same time, a few pilots proved to be unusually adept at shooting down the enemy.

These aces represented only about one in twenty of all pilots, yet accounted for the majority of kills. Among the most successful of them was the Canadian, Raymond Collishaw, of the Royal Naval Air Service (RNAS), and who later served in the Royal Air Force (RAF).

## IN AT THE DEEP END

Collishaw was born in Nanaimo, British Columbia, in 1893, and started work aged fifteen as a cabin boy on a Canadian

**7** Raymond Collishaw in RAF uniform (1919).

Fisheries Protection Service ship. Gradually he worked his way up the ranks and at the start of the First World War tried to enlist in the Royal Canadian Navy. When he got no reply he applied instead to the Royal Naval Air Service (RNAS) and eventually ended up at Redcar, in England, to do what in those days passed as flight training. Aircraft simply hadn't been around that long in 1916, and learning how to fly was a pretty haphazard and perfunctory affair.

Part of Collishaw's training was in a French-made Caudron G.3, which was so primitive that it used the old Wright brothers' patented method of wing warping to control the plane's roll (rotation about the nose-to-tail axis). Wing warping involved a system of levers and pulleys to twist the trailing edges of the wings in opposite directions, in the same way that the flight of a paper plane can be adjusted by curling the back edges of its wings. In most aircraft – biplanes and monoplanes – after 1915, ailerons had taken the place of wing warping to control banking and rolling.

Despite some trouble with his landings, Collishaw flew solo after only eight and a half hours of flying time. That was about par for the course on both sides during the First World War. Flight instructors showed a novice pilot how to get in the air, turn the plane around, and land again, and not much more. Ground instruction consisted of learning how to load and unjam a machine gun and basic map reading. Everything else of use to a pilot in battle had to be learned on the job, more often than not with an enemy plane trying to perforate your fuselage. No wonder so many young military pilots at the time were lost in training and operations, and why those who survived and became skilled were able to notch up so many kills.

**8** Royal Flying Corps or Royal Air Force Sopwith 1 ½ Strutters.

Collishaw was luckier than most. Having been posted to the front with hardly any flying experience, he was mentored by none other than John Alcock, who, in 1919, along with Arthur Brown, would make the first non-stop flight across the Atlantic in a converted Vickers Vimy bomber. Thanks to Alcock's expert coaching, Collishaw soon became a better pilot than most of his contemporaries. Even so, he made his fair share of mistakes. On one occasion, while on an unauthorized detour trying to deliver a message from a friend to a local girl, he crashed into a row of outhouses, covering himself in a variety of unappealing materials and wrecking the plane. Needless to say, the object of adoration was less than impressed.

Having earned his wings he was posted to Naval 3 Wing, which had just started flying the new Sopwith 1 ½ Strutters, both as one-seater bombers and two-seater fighter scouts from Luxeuil-les-Bains, France. Fitted with special fuel tanks, they were able to carry out long-distance raids over Germany. One of Collishaw's first missions, piloting a two-seater as cover for the bombers, was a strike on the Mauser arms works at Oberndorf. More than eighty aircraft took part in the raid – a lot by 1916 standards – but unfortunately the contingent of Nieuport 11 scout planes that started out with the party had to turn back when they ran short of fuel, leaving the rest of the armada vulnerable to attack by German fighters. Sure enough, having crossed the Rhine, the allied raiders were set upon by a group of Albatros DIIs. The mission was disrupted, but that day saw Collishaw shoot down his first enemy plane and a prize scalp to boot – that of the future ace Ludwig Hanstein. His gunner having riddled Hanstein's engine with bullets, Collishaw followed him down, firing at the German with his forward Vickers. Shortly after, in another raid, Collishaw's plane was hit, but fortunately the light, big-winged aircraft of that era could glide well, and he was able to coast into Allied France near Nancy and hard land without injury.

## MEET THE FOKKERS

Come January 1917, Collishaw was delivering a Strutter to its new base at Ochey, without the protection of a rear gunner, when he strayed over the front line and was attacked by half a dozen Albatros DIIs. He learned he had company when bullets started making a mess of his instrument panel, one ricocheting off his goggles and partially blinding him with splintered glass.

Instinctively he dived low, hoping to shake off the pursuers. One German plane followed and crashed; another cut in front of him and was dispatched with a well-aimed burst of machine-gun fire, after which the rest broke away. But now Collishaw had to find his way home without instruments and with blood running into his eyes.

Using the sun as a guide, he headed in what he thought was roughly the right direction and was happy to touch down on the nearest usable airstrip. His relief at seeing a crew running towards him quickly evaporated though when he noticed a line of Fokker E.1's – single-winged fighter planes that were the scourge of Allied aviators on the Western Front: he had landed at a German base. Keen to avoid the reception commit-tee, he speedily took off again, some of the Fokkers in pursuit, and shaved a few branches off the trees at the end of the field as he struggled to climb. The enemy planes were faster and closed on him, peppering him with bullets, but he managed to gain enough height to lose them in clouds. He'd flown several miles into friendly territory before he got his bearings and was able to touch down at a French airfield near Verdun. There he spent a number of days recuperating and having his eyes treated by a local doctor.

So impressed were the French by his exploits that they awarded him the Croix de Guerre, while the British, recogniz-ing they had a rising aerial star on their hands, sent him to join an all-fighter squadron. From Collishaw's point of view, this was like being tossed from the frying pan into the fire. It was the spring of 1917, and the Allied squadrons on the Western Front were pummelling and being pummelled as part of the Battle of Arras – a British-led offensive against German defences near the city of Arras in northern France. Several groups of

# Sopwith Triplane

**Crew:** 1
**Length:** 18 ft 10 in (5.73 m)
**Wingspan:** 26ft 6 in (8 m)
**Empty Weight:**
1,101 lbs (500 kgs)
**Maximum Speed:**
117 mph (187 km/h)
**Powerplant:**
Clerget 9B rotary engine, 130 hp
**Armament:**
1x .303 inch Vickers Machine gun

**9** Sopwith Triplane.

naval fliers were brought in to support the effort but, for the most part, these men lacked battle experience and found the demands of almost continuous dogfighting, often involving several flights a day, too much for them. On more than one occasion, Collishaw suddenly discovered himself alone with the enemy, the rest of his RNAS colleagues having headed back to base due to a variety of mysterious mechanical problems. In one particular gnarly encounter his goggles were again shot off and, to maker matters worse, his gun seized up so that he had to lean into the slipstream without eye protection to unjam the weapon. A month in hospital followed, recovering from a nasty case of frostbite.

## BLACK FLIGHT

Back on the front in April 1917, Collishaw was posted to Naval 10 Wing, which had just received a batch of brand-new Sopwith Triplanes. Although a bit slower than the opposition's Albatros DIII, the Triplane was a lot more nimble, with a small turning circle and a spectacular rate of climb. Its agility more than made up for the fact that it carried only a single machine gun against the two of the Albatros, and the Germans came in for a shock when they confronted it. On Collishaw's first day piloting the Triplane, he downed a plane, and added four more to his tally over the next few weeks.

Naval 10 was then moved to Droogland, near the Belgian border, in preparation for the Messines offensive in early June. The Royal Flying Corps needed extra fighter cover for its reconnaissance and bombing missions. Now Collishaw and the rest of Naval 10 would be tested against the cream of the

German Army Air Service, including Baron von Richthofen's infamous 'Flying Circus'. Each flight in the Naval 10 squadron painted the cowlings of their Triplanes a different colour so they could be easily identified in the air. Collishaw commanded 'B' flight, consisting of five pilots, all Canadian, and their colour was black – not just on the cowling, but on the whole aircraft. To complete the livery, each pilot chose a moniker that was also painted on the plane. There was 'Black Prince' and 'Black Death', for instance, and Collishaw's own 'Black Maria'. 'B' flight became known as 'Black Flight' and pretty soon they were the scourge of the German air force.

On the eve of the Battle of Messines came one of Black Flight's finest hours. While on offensive patrol, they ran into a large group of Albatros and Halberstadt fighters. Collishaw brought down three of them, and his colleagues seven others, with no loss to themselves. Then, two days later, Collishaw had an astonishing escape. While engaging an Albatros he was hit from behind by another aircraft. His Triplane started to fall in a series of wild cartwheels, swoops, spins, and dives, plunging 16,000 feet. Just before it struck the ground, Collishaw managed to pull up the nose of the plane and bellyflop down. A group of English foot soldiers pulled him from the wreckage, dazed but otherwise in good shape.

A month later, he was shot down again. This time, a hail of bullets from a German aircraft snapped the wires that held together the two halves of his engine cowling. One half flew off and slammed into his wing struts, causing the plane to go out of control and tumble end over end. The extra force on his seat belt made it snap and Collishaw was thrown out of the cockpit. Somehow he managed to grab hold of a strut and hang on, but as he was tossed this way and that like a rag doll, he felt his grip loosening. Luckily a gyration of the aircraft threw him part-way back into the cockpit, and he was able to hook a boot around the control column and push it forward enough so that he could get further in. Almost at tree level, with one last desperate effort he levered himself back into the cockpit and yanked back on the stick. It was enough to save his life. The Triplane hit the ground and was a write-off. But Collishaw had levelled out enough that he was able to walk from the scene of destruction – once again without injury.

## ENCOUNTERS WITH THE RED BARON

Black Flight had already accounted for fifty enemy planes when, on 27 June 1917, they engaged von Richthofen's No. 11 fighter squadron. One of the Blacks, Gerry Nash, found himself single-handedly taking on two German planes – but not just any planes. One was piloted by Karl Allmenröder, a cool veteran ace with some thirty kills to his name, the other by the Red Baron himself, von Richthofen. Nash gave these deadliest of foes a tremendous run for their money, but finally Allmenröder got in a machine-gun burst that took out some of the Sopwith's instrument panel. Nash managed to land, but it was behind enemy lines and he was taken prisoner.

The four remaining members of Black Flight thought that Nash had been killed and swore to avenge him. Later the same day, they ran into Richthofen's squadron again, near Courtrai, and this time it was Collishaw who was pitted against the bright green Albatros of Allmenröder. What happened next isn't clear. According to some accounts, Collishaw shot down

Allmenröder with a lucky, long-range burst, but the Canadian staked no such claim in the records for that day. Allmenröder was certainly killed on 27 June but exactly how and by whom remains a mystery. Nash, lying in a cell, heard a church bell tolling some days later, and learned from his guard that it was for the funeral of the German ace.

On 6 July, Collishaw was involved in another air battle with von Richthofen and his men, which almost saw the end of the Red Baron. A small group of Royal Air Corps FE2b pusher biplanes had come under attack by about thirty Albatrosses, including the Red Baron and seven others from his squadron, when Black Flight joined the fray. In the ensuing melee, Collishaw shot down no fewer than six enemy scout planes – his personal best for a single day. Meanwhile, a gunner on one of the FE2bs grazed von Richthofen's skull with a bullet, temporarily blinding and disorientating the German ace. The Baron's plane went into a spin but he came to in order to gain sufficient control and land heavily in a field. A head wound put him out of action for a month, requiring extensive surgery to remove splintered bone, and, some medical evidence suggests, left him with permanent brain damage.

The day after Collishaw's titanic encounter he was back in the air again, such was the pressing need for experienced pilots. But the end of the month saw the disbandment of Black Flight, which collectively had dispatched eighty-seven German planes with the loss of only two of its own men. Collishaw was allowed to return to Canada for three months to recover from combat stress.

## STING IN THE TAIL

On returning to Europe Collishaw was posted to Naval 3 squadron, which was flying a hot new fighter plane called the Sopwith Camel F1. Shortly after he arrived, the squadron leader was killed and Collishaw was given command. By all accounts, he was brilliant at the job – charismatic and inspirational. Often he would escort rookie pilots and let them go in first against enemy planes and then, if they failed to make the kill, fly in behind them and apply the *coup de grâce*. Back on the ground, he would congratulate them on their first victory.

Collishaw's adventures, both in the air and on the ground, didn't end after the Great War. In 1919, now a member of an RAF squadron, he was involved in setting up the Baghdad-to-Cairo air route to enable easy military access between Britain and what was then called Mesopotamia. The pilots worked with two truck crews on the ground to identify landing areas, roughly twenty miles apart, preferably near water and on flat ground. At one of the oases where the aircraft landed they needed to take on water, but it was at the bottom of an eighty-foot well and no one was keen to go down and get it. In the end, Collishaw himself went down on the end of a rope, bucket in hand. By the time he reached the bottom his eyes had adjusted to the dark, and he was horrified to see the walls above him crawling with huge black scorpions. Quickly he scooped up a bucket of water and rose out of the well. One bucket was all they got.

# 4

## DANCES WITH DEATH

What could you do, back in the 1920s, if you had a plane to spare, loved flying and extreme thrills, and wanted to earn a living at the same time? The answer is you could barnstorm. After the First World War a glut of aircraft came on to the market at bargain-basement prices. The US military was trying to offload a lot of the Curtiss JN-4 biplanes, known as Jennys, which had taken part in the conflict and were now surplus to requirements. Costing over $5,000 new, a Jenny could sometimes be found on sale for as little as $200. A good number of military pilots who had become expert at flying the Jennys in battle were now keen to snap up the planes for their own civilian use. Plenty more low-cost planes, some of whose designs were ahead of their time, were also going cheap as their manufacturers went bust, due to the fact that the civil aviation industry hadn't mushroomed as quickly as some had hoped.

During the early 1920s, there were hundreds of talented American pilots who were prepared to fly their planes anywhere, and do whatever they could to make money. Some delivered mail; a few carried more illicit cargo, such as carrying whiskey across the Mexican border. But of all the flying activities, barnstorming was the most popular and profitable

**Crew:** 2
**Length:** 27 ft 4 in (8.33 m)
**Wingspan:** 43 ft 7 in (13.3 m)
**Empty weight:** 1,390 lb (630 kgs)
**Maximum speed:** 75 mph (121 km/h)
**Powerplant:** Curtiss OX-5 V8, 90 hp

**10** The Curtiss JN-4 ('Jenny').

for those prepared to take the risks. It was encouraged too by the lack of federal regulations on aviation, which allowed all kinds of crazy stunts to be indulged in, without much thought to the safety of those either in the air or on the ground.

Many young fliers who had returned home to America from Europe after the First World War took up a drifter lifestyle, drifting from town to town across America, offering rides for money, meals, or gasoline, and sleeping out in fields under the wings of their planes. They had to be jacks of all trades, including self-taught engineers, because no one else was around to fix their machines if they broke down or needed spare parts.

Aircraft were still a comparative novelty in those days and, especially in rural areas where there was plenty of space to fly in and few other attractions, people came in droves to watch the latest hair-raising aerobatic display. Not surprisingly, the most successful barnstormers were the ones who put on the most spectacular shows and, just as importantly, had a talented promoter working alongside them to pull in the crowds. One such star of the barnstorming era was Ormer Leslie ('Lock') Locklear.

## BORN TO BE WILD

Born in 1891, one of ten children, Locklear spent most of his early life in Fort Worth, Texas, and got a grounding in carpentry, his dad's profession. Even as a school kid, though, it was obvious he had far too much adrenaline for a quiet life of woodworking and construction. He proved he had a devil-may-care attitude towards his personal safety by doing wild stunts on his bicycle – pedalling furiously up a plank ramp onto a barn roof and then leaping off, Evel Knievel style, onto a platform below. His passion for aviation started when he saw his first planes.

Early in 1911, a team of international fliers, mostly French, had been putting on aerial displays in nearby Dallas. A group of businessmen persuaded them to travel the thirty miles or so to Fort Worth. For the two days of their show, stores closed, schools announced half-day holidays, and cheap streetcar rides were offered to the site of the big event. Ormer Locklear and some of his friends were among the 15,000-strong crowd that looked on, dazzled by the exhibition of piloting skills. In that same year, the pioneer flier Calbraith Rodgers landed at Fort Worth to fix a clogged fuel line during the first transcontinental flight across North America. Locklear was again watching, this time on the Hattie Street Bridge with his brothers.

Not having access to a powered aircraft, but inspired by what they had seen, the Locklear boys built a glider with fifteen-foot wings from bamboo fishing poles and shellacked linen. They pulled this behind a car so that it rose up to 150 feet in the air, each brother taking a turn at flying the fragile craft back to the ground, guiding it the only way they could – by shifting their body weight from side to side. In search of further thrills, Locklear graduated to a two-cylinder Indian motorbike and sped around his neighbourhood, popping wheelies and even standing on his head while steering the machine. On one occasion he shot across a busy street in downtown Fort Worth and ended up plunging through the doors of a saloon.

News of his exploits started to get around. The famous magician and escape artist Harry Houdini learned of them when he came to Fort Worth and did a series of shows at the Majestic Theatre. One day he struck up a conversation with Locklear's younger brother in the sporting-goods store where he worked. Houdini reasoned there might be some good publicity to be had by teaming up with this local stunt kid, and was completely sold on the idea when he met the charismatic and good-looking Ormer – a crowd-pleaser if ever there was one.

The great escapologist proposed that Locklear drag him, hog-tied, behind his motorcycle down Main Street in Fort Worth. While hurtling down the road, with wide-eyed spectators lining the way, he would break out of his bonds and roll free. The young man, not usually averse to risk, wasn't keen on the idea but Houdini won him round by insisting that he would take every precaution so as not to be injured. He chose Main Street because it was the only one in town that was paved, if wooden posts set into the ground to protect horses' hooves counted as paving. To protect his body and legs he would wear thick, quilted overalls and have a hood round his head.

Locklear agreed, and the next day sat astride his bike while Houdini had his hands tied behind his back by volunteers and then lay down in the street. Revving his machine Locklear rolled forward, taking up the slack on the rope that was knotted around Houdini's ankles, before opening the throttle and roaring away. Evidently, the stunt was a success because the magician went on to play the rest of his scheduled shows in Fort Worth – to packed audiences.

## IN THE AIR AT LAST

In April 1917 the United States entered the First World War and on 25 October of that year, a few days short of Locklear's twenty-sixth birthday, he enlisted in the Army Air Service. To celebrate, he planned to ride his bike on the wall around the top of a new building in Fort Worth and, the night before, set up platforms so that he could take the corners more easily. He told the press of his intentions, and a few hundred locals gathered the next day to watch him. His plans were stymied at the last moment, however, because of a complaint by the Humane Society.

Any disappointment on Locklear's part, though, was quickly forgotten. After ground training at the Texas School of Military Aeronautics he was assigned, with the rank of second lieutenant, to Barron Field near Fort Worth to finish his flight training, and now finally he got his hands on a real-life gasoline-powered aircraft. He flew a Canadian version of the Curtiss Jenny, with a ninety-horsepower engine and a top speed of ninety miles per hour, and right from the start he showed his true colours.

Flying seemed to come as easily to him as breathing, and before long, fearless as ever, he would occasionally clamber out of his pilot's seat onto the wing or over the cowling, thousands of feet above the ground. The first time he did this was during a test in which the aim was to read ten words on the ground while flying at 5,000 feet. No one had ever achieved a perfect score because the plane's lower wing always got in the way. Locklear's solution was to climb onto the wing and look down in front of it while his instructor kept the aircraft on an even keel. He got all the words right – and everyone else who tried to copy his trick afterwards was reprimanded for their trouble.

On another occasion, Locklear was in the front seat of the Jenny when a radiator cap worked its way loose causing hot water to spray in his face. He climbed out of his seat, inching forward over the fuselage until he could reach the cap on the exposed engine and screw it back on. Dangling radiator caps, loose spark-plug wires – whatever it happened to be – the young flier thought nothing of unbuckling his safety belt and sorting out the problem right on the spot, in mid-air.

Ormer Locklear may not have been the first person ever to do 'wing walking' but he was the first to make a regular habit of it and, eventually, to turn it into a lucrative profession. One day he set out to prove his superiors wrong in their belief that no extra weight could be put onto the leading edge of an aircraft's wing without seriously disrupting the plane's aerodynamics. With a student at the controls, Locklear crawled out of his cockpit onto the Jenny's lower wing and made his way to stand on the leading edge. Lo and behold, nothing untoward happened – an important discovery, from the military point of view, because it showed that guns could be safely mounted on the front of a wing away from the fuselage.

Pretty soon, Locklear and a couple of like-minded friends, Milton 'Skeets' Elliot and James Frew, were doing aerial stunts on a regular basis whenever they got the chance. Word reached the commanding officer of Barron Field, Colonel Thomas Turner, and he went out to see the three men in action. Taking

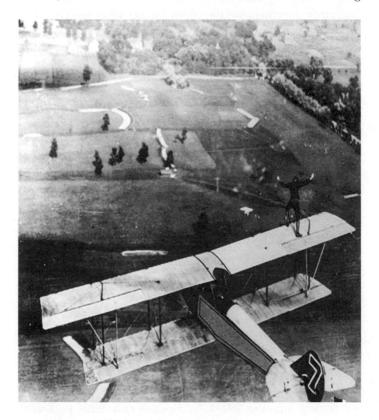

**11** Ormer Locklear 'wing walking', *c.*1919.

the smart option, instead of dressing down the reprobates he invited them to put on official demonstrations to show off the agility and stability of the Jenny. Their piloting skills led to their remaining at Barron Field as instructors instead of being shipped out to fight in France where, if they had survived, they would probably have become among the top aces of the war.

Not surprisingly, given that their stunting was now officially sanctioned, Locklear and his buddies pushed their luck to the limit. On 8 November 1918, Locklear made history by transferring between planes in mid-air, dropping from the undercarriage of Frew's plane onto the wing of Elliot's, which was flying directly below. Within months he had perfected a different kind of plane-to-plane transfer that was even riskier.

## KING OF THE WING WALKERS

Before leaving the Army Air Service in May 1919, Locklear met William Pickens, an experienced press agent who had already been successful in promoting Lincoln Beachey and some other outstanding young fliers. Locklear, Skeets Elliot, and another stunt-flying instructor from Barron Field, Shirley Short, signed a contract with Pickens to appear in aerobatic shows around the country. Within a few months, Locklear had achieved national and international fame as the 'King of the Wing Walkers'.

The skills he had honed while wing walking for a serious purpose in the military, he now put before the public for entertainment. Together with Elliot and Short, he developed new tricks, including various ways of hanging from the lower wing by a rope ladder or trapeze bar, or, on the upper wing, performing handstands. It so happened that the Curtiss Jenny came as standard with over-wing struts that served ideally as braces for wing walking, although spectators who had never seen them before often assumed they had been added specially for doing stunts.

Of all Locklear's aerial antics, none was more impressive and dangerous than his 'Dance of Death' – a manoeuvre so incredibly risky that it has never been attempted since. It involved Locklear flying one plane while Elliot flew another, right alongside, so that the wings of the two aircraft were almost touching. On a signal, the two pilots would lock their controls in place, climb out of their cockpits, scamper along the wings – passing each other along the way – and then hop into the cockpit of the other plane. Such audacity and breathtaking skill drew in huge crowds, thanks also to the deft handling of the media by Pickens. Soon Locklear grew wealthy from his escapades, earning up to $3,000 a day – the equivalent of about $50,000 today. This was in stark contrast to the hand-to-mouth existence of most barnstormers. But even Locklear, aerial megastar that he was, continually had to dream up new ideas and madcap exploits to keep his public interested.

It wasn't long after Locklear became a global celebrity that Hollywood took an interest in him. One of his aerobatic displays in Los Angeles caught the attention of the movie-making community and was followed by a number of well-publicized exhibitions at an airfield owned by Sydney Chaplin, brother of the famous comic actor. In no time, Pickens had arranged for Locklear to appear as a stuntman in Universal's film *The Great Air Robbery* (1919). During the shooting, in July 1919, Locklear made the first successful transfer between a moving plane and a speeding automobile below. In a later shot he did one of his trademark plane-to-plane hops, before rounding off his first venture into Hollywood escapism with another plane-to-car switch, in which he dropped into a car to wrestle with a villain before grabbing onto the undercarriage of the plane just feet above him and climbing away to safety, moments before the car flipped over and crashed in a ball of flames.

## FALLING STAR

Shortly after Locklear arrived in Hollywood, he met the vivacious young actress Viola Dana. At twenty-three she was already the star of dozens of films stretching back to her childhood, and was Buster Keaton's off-screen leading lady. Locklear and Viola began a very public romance, becoming one of the celebrity couples of the time. They even, according to local press reports, became engaged, despite the fact that Locklear already had a wife, Ruby, from whom he was separated, living back in Texas. Viola worked at Metro Pictures (which later became part of MGM), and her high-flying lover would often buzz her on the Metro lot in his Jenny, sometimes ricocheting off the roof of a soundstage where he thought she might be filming.

In December 1919, while awaiting the premiere of *The Great Air Robbery*, Locklear and his flying partners put on a series of aerial shows in California. One in San Francisco, on the 19th, was a benefit performance – and for this Locklear pulled out all the stops. Hanging one-handed from a skid, he waved to the crowd while the plane flew so low that onlookers could see him grinning. Then he clambered onto the upper wing, made his way to the tip, and posed while balanced on the ball of one foot.

Soon Locklear was basking in the glow of Hollywood glamour, surrounded by titans of the industry such as Cecil B. DeMille (an aviation enthusiast himself and owner of an airfield), Charlie Chaplin, Douglas Fairbanks, and Rudolph Valentino. In April 1920, the William Fox Studio (now Twentieth Century Fox) signed him to star in *The Skywayman*, on a wage of $1,650 a week, alongside the Australian movie queen Louise Lovely. As the title suggests, the film revolved around furious aerial-action sequences. There were two plane-to-plane transfers: one done by Skeets Elliot and a second by Locklear himself. This was followed by a gunfight in which Locklear perched on the landing gear of a plane in flight, while the crooks raced away in their car.

The movie's climax involved a night-time action scene over an oilfield, which the director had wanted to shoot during the day and then darken with the help of filters, or use miniatures instead. But Locklear insisted on doing the whole thing live, and at night. Other pilots were successfully copying his stunts and coming up with new ones, and he wanted to prove that he was still ahead of the game. So, against the advice of the studio, on 2 August, the very last day of filming, he persuaded Elliot and Short to do the night-time shoot.

Searchlights lit up the skies above an oilfield next to DeMille Airfield. The plan was for the lights to illuminate the plane carrying Locklear and his pilot Elliot as it made a spiralling dive of more than 4,500 feet towards the oil derricks, then to switch off at the last minute to allow the plane to level off. Tragically, the lights stayed on, blinding Elliot and causing him to smash into a pool of oil next to the well, and the plane to erupt in a fireball. Both men died instantly. Somewhat gruesomely, Fox included the crash and its aftermath in the final release, together with a closing scene – shot earlier – in which Locklear and Elliot are seen walking safely away from the accident. Among those watching the horrific events of the night unfold was Viola Dana, who afterwards refused to fly for more than a quarter of a century.

Locklear's funeral ceremonies – two of them, in Los Angeles and Fort Worth – would have been worthy of the greatest of movie icons. Thousands of fans gathered outside the funeral home where the aviator's glamorous girlfriend wept over his coffin. In Texas, his estranged wife did the same, before Locklear and Elliot were both buried in their hometown. Years later, in *The Great Waldo Pepper* (1975), which Robert Redford directed and starred in, the life story of the main character was loosely based on that of Locklear. At its premiere, one of the honoured guests was Viola Dana.

# 5

## UNDER PRESSURE

The first pilot to fly solo around the world did it in under eight days on less than twenty hours' sleep. To warn himself if he nodded off while at the controls he tied a heavy tool by a string to one finger so that if the tool fell he would be jerked instantly awake. On one occasion he almost ran out of fuel and considered bailing out; at other times, the weather was so bad he couldn't see where he was going. And just to make it that bit harder, he had only one eye.

Wiley Post was born in Texas in 1898 and spent most of his youth in rural Oklahoma. His parents were farmers but, as a youngster, Post had no desire to go down that route; his big interest was in anything mechanical. In 1913 at the county fair in Lawton, he saw his first aircraft in flight – one of the original Curtiss 'pusher' planes that looked like an oversized box kite with a big propeller at the back. There and then he knew what he wanted to do, and began to pick up the basic knowledge in science, maths and navigation that would help him in his goal to become a pilot. But his path into aviation turned out to be anything but quick and easy.

During the First World War, Post joined the US Army with the dream of flying but ended up instead as a radio operator. On returning home he took a job as a roughneck in the Oklahoma oilfields and began what was to prove to be a troubled time for him. In 1921 he fell foul of the law after being convicted of armed robbery (of a car) and was sentenced to ten years in the Oklahoma State Reformatory, although he was paroled after just thirteen months.

Post's career in flight didn't really kick off until 1924 when he joined an aerobatics group – Burell Tibbs and His Texas Topnotch Fliers – not as a pilot but as a stunt parachutist. It was during this time that he learned how to fly, from experienced pilots in the show. After that he decided he had to have a plane of his own, something that was only going to happen if he earned a heck of a lot more money than the flying circus was paying him. So, in 1926, he went back to the oilfields, only to be injured on his very first day: a piece of metal flew into his left eye, robbing him of its sight. An infection from the injury threatened to blind him completely, so he agreed to surgeons removing the damaged eye. For the rest of his life he wore a patch, making him instantly recognizable. But the accident proved to be a blessing in disguise because with the $1,800 settlement he received, he was able to buy his first plane – a Canadian-built version of the Curtiss JN-4, called a Canuck.

## WHEN WILEY MET WINNIE

Around this time, Post crossed paths with fellow Oklahoman Will Rogers, humorist, star of the silver screen in the 1920s and 1930s, and all-American cowboy. Rogers needed to get to a rodeo, and Post was happy to fly him there. The two got along and Rogers introduced him to a friend of his,

Florence C. Hall, a wealthy oil magnate. That meeting, in 1927, led to Post becoming Hall's personal pilot and having use of his plane, an open-cockpit Travel Air biplane. A year later, Hall decided he wanted to upgrade to a newer, closed-cockpit model and asked Post to visit Lockheed's factory in Burbank, California, and pick out the best the company had to offer. Post chose a streamlined, high-wing, Lockheed Vega monoplane, which the oilman promptly named *Winnie Mae*, after his daughter.

Hall's business took a hit during the Great Depression, and he had to dispense temporarily with both his aircraft and its pilot. Post worked for a while with Lockheed as a salesman and test pilot. But in 1930 Hall was back in the black and able to reinvest in a newer version of the Vega, again nicknamed *Winnie Mae*, and give Post his old job back. Not only that, but Post was encouraged to forge ahead with one of his long-standing

**12** Wiley Post (unusually not wearing his eye patch) and Harold Gatty.

ambitions – to fly around the world in record-breaking time. As a prelude to this, Post entered the National Air Race Derby from Los Angeles to Chicago. For the competition, he had Lockheed install a new 500-horsepower engine into the *Winnie Mae*, which pushed the plane's top speed up to about 200 miles per hour. In the souped-up Vega he took first place, making the trip in nine hours eight minutes and thrusting him onto the world stage.

Like many aviators, Post wasn't happy about the fact that the record for flying around the world was held by an airship, not a plane. Hugo Eckener's twenty-one-day circumnavigation of the globe in the *Graf Zeppelin* had been achieved the previous year. To help Post in his quest to smash this time, he hired the Australian Harold Gatty, himself a veteran of several trans-continental flights, as his navigator. Gatty was also a technical genius, having developed several new way-finding devices, including a combined ground-speed and wind-drift indicator. This instrument was particularly useful to Post, because having only one eye meant he lacked depth perception, which made it hard for him to gauge distances and speed.

On 23 June 1931, Post and Gatty took off from Long Island heading north and then west, landing in Newfoundland to refuel, before continuing across the Atlantic and on to Europe, making further stops in England, Germany, and the Soviet Union. At an airfield in Siberia, they hit their first major snag. Torrential rain left the *Winnie Mae* bogged down in mud and it was only after wasting fourteen hours struggling to free the plane that the aviators were helped by a detachment of American soldiers with a tractor. Several hours more were spent on the ground in Khabarovsk while mechanics checked over the engine in preparation for the Vega to attempt the longest leg of its journey, across the Pacific. After a seventeen-hour flight, Post and Gatty landed in Solomon, Alaska, and then encountered their most serious setback – one that threatened the entire venture. In their book about the journey, *Around the World in 8 Days*, Post described the almost ill-fated take-off from Solomon:

> Taxiing back along the beach, the ship started to sink into the sand. With a quick thrust I banged the throttle open to pull her through it before we were stuck. But all I succeeded in doing was to boost the tail up into the air. With a loud slap the propeller cut a hole in the sand and bent both tips on the blades. I cut the emergency switch just in time to keep 'Winnie Mae' from making an exhibition of herself by standing on her nose. That would have been fatal to our hopes.
>
> I jumped out and surveyed the damage. With a wrench, a broken-handled hammer, and a round stone, I drew out the tips of the blades so they would at least fan the air in the right direction.
>
> Harold was swinging the prop for a prime with the switch cut to restart the hot engine. He called 'all clear' to me, and I switched on and whirled the booster. One of the hot charges of gasoline caught on the upstroke of the piston, and with a back fire the Wasp [engine] kicked. The propeller flew out of Harold's hands, and the blade opposite smacked his shoulder before he could jump clear of the track. He dropped like a log. It was fortunate, to say the least, that it was the flat side of the blade which hit him,

though it gave him a bad bruise and a wrenched back. If the prop had been going the other way, he might have been sliced in two.

Safely back in the air, *Winnie Mae* headed for Fairbanks to have her prop replaced with a spare obtained from Alaska Airways. Next was a climb over the 9,000-feet-high Rockies to Edmonton, where the plane landed on another strip so water-logged that it couldn't be used for take-off. Instead, with the help of locals, the plane was dragged over to a street in the city that served as a usable runway, although the wings had only a few inches clearance of obstacles on either side. After a final refuelling stop in Cleveland, Post and Gatty arrived back at Roosevelt Field in New York, from where they had started the adventure eight days, fifteen hours, and fifty-one minutes earlier, having covered 15,500 miles and shattered the old world record.

**13** The *Winnie Mae* on display in the National Air and Space Museum.

## GOING IT ALONE

A rapturous welcome home followed, including lunch at the White House and, the next day, a 'ticker-tape parade' in New York City, topped off by a banquet for the two heroes hosted by the Aeronautical Chamber of Commerce of America at the Hotel Astor. To complete the triumphant return, Florence Hall gave Post the *Winnie Mae* as a gift in recognition of his feat.

But in the aftermath, Post was left unfulfilled. The press started to credit the success of the record-breaking flight more to Gatty's technical know-how than Post's piloting skills. And when Post tried to push ahead with plans to start his own aeronautical school, he found financial backers thin on the ground because of doubts about his rustic background and lack of formal schooling. He also realized that other fliers would soon try to break the new record and he wanted to set the bar so high that it would be beyond the reach of anyone for years to come. Motivated by these factors, he set out his next personal challenge: to beat his own world record – and this time to do it on his own.

Over the next year or so, Post worked on getting *Winnie Mae* in shape for the challenge. Without a navigator he would have to take readings and figure out his position and heading at the same time as flying the plane. To help him do this he installed an autopilot and a radio direction-finder that the Sperry Gyroscope Company and the US Army had been prototyping. Oddly enough, Post's eye injury helped him in his challenge, because he was used to doing off-the-top-of-his-head calculations to make up for his lack of depth perception. (He often kidded that he would have to give up flying if ever they changed the height of two-storey buildings.) As he had for his flight with Gatty, Post trained physically and mentally for the ordeal, getting used, for example, to taking short naps instead of sleeping for hours at a stretch.

On 15 July 1933, he set out from Floyd Bennett Field in Brooklyn, New York, and didn't stop until he reached Berlin, twenty-six hours later. Following a similar route to his earlier flight with Gatty, he landed several times in the Soviet Union for refuelling and repairs, especially to the autopilot, which kept acting up. Fighting the urge to fall asleep, exhausted from the mental and physical demands of days of solo flying over vast distances with only the soporific drone of the engine for company, he finally arrived back over North America, only to have his radio direction-finder go awry. Worried about the tall mountains in his way, he touched down on a tiny landing strip in the isolated mining town of Flat, damaging his propeller and right-side landing gear in the process. After repairs, with the help of some local miners, he flew on to Edmonton, and then 2,200 miles non-stop to New York, where, waiting for him at Floyd Bennett Field, was a crowd of 50,000 people. He had completed his solo flight around the world in just seven days nineteen hours, knocking twenty-one hours off his previous record – a feat that, given the obstacles he had faced and the technical limitations in those days, was astonishing.

Post next considered entering the race for the £10,000 Robertson Prize, between England and Australia, to be held the following year. But he knew *Winnie Mae* wasn't as fast as some of the newer planes that would be taking part. His only chance would be to fly very high, between 27,000 and 36,000 feet, where earlier experiments with balloons had suggested there were rapid currents of air (what we now know as the jet stream). To survive and function properly at such altitudes a pilot would need to breathe pressurized air. Since *Winnie Mae*'s airframe wasn't strong enough for the whole cabin to be pressurized, the only solution would be to wear some kind of special suit similar to the ones used by deep-sea divers.

## DRESSED FOR ALTITUDE

Post was keen to keep his high-flying plans secret, but he did talk about the pressure-suit idea with his aviator friend Jimmy Doolittle, who referred him to the B.F. Goodrich Company. Post asked the firm if they would help him develop a suit and helmet that, when the plane was at altitudes of up to 27,000 feet, would maintain a pressure equivalent to being at 5,500 feet – the air pressure that would be provided by a new supercharger that he planned to install on *Winnie Mae*'s engine. Goodrich ended up making three suits of rubberized parachute fabric, each one more refined than the last. The third one passed all the pressure tests it was subjected to and was flexible enough at the knee and elbow joints to allow good freedom of movement even when inflated.

In late 1934, Post flew several test flights wearing the suit, up to 40,000 feet. On 7 December, he reached 50,000 feet and actually rode the jet stream, but lost the chance to set an official world altitude record when one of the two barographs (used to measure altitude) in the plane failed during the ascent.

The final version of the pressure suit and the upgrades to *Winnie Mae* didn't come in time for the Robertson Prize race, but Post carried on with his high-altitude experiments regardless. An attempt at a transcontinental speed record could have ended in disaster when a jealous pilot sabotaged *Winnie Mae*'s engine. And a subsequent transcontinental flight had to be aborted when Post ran out of oxygen over the Midwest, but he did manage to fly from Burbank to Cleveland in a record time of seven hours nineteen minutes at an average speed of 279 miles per hour – a full 100 miles per hour faster than the plane's normal maximum airspeed – thanks to a massive boost from the jet stream.

The old plane was getting too long in the tooth to keep up this kind of workload, and its breakdowns became more and more frequent. After four unsuccessful transcontinental flights at high altitude, *Winnie Mae* was retired and, in 1935, sold along with its original instruments to the Smithsonian Institution. Today, it can be seen on display in the National Air and Space Museum in Washington, DC.

## DEADLY COCKTAIL

Post became interested in setting up a passenger and airmail service between the US West Coast and Russia. Not having enough funds to buy a new plane for the job, he decided to build a hybrid from parts salvaged from two other aircraft. The low-wing monoplane would have the fuselage of a Lockheed Orion and the long wings of an experimental Lockheed Explorer that had crashed. Into this airframe, Post installed a 550-horsepower Wasp engine and oversize 260-gallon gas tanks. The final step would be to add pontoons so that the plane could take off from, and land on, the plentiful lakes in Alaska and Siberia.

Lockheed refused to let their engineers have anything to do with the project, on the grounds that fusing together two different designs could produce a dangerous mix – a position that was soon to be tragically vindicated. Post forged ahead nevertheless, assembling the Orion-Explorer cross-breed, or what others less charitably called 'Wiley's Orphan' or 'Wiley's Bastard', with the help of Pacific Airmotive Ltd at the Lockheed airport in Burbank. Will Rogers often visited him there to check on his progress and asked Post if he would fly him to Alaska when the new plane was ready, to collect material for his newspaper column. The plan was for Post and his wife, Mae, to fly up to Seattle to have the pontoons fitted, meet Rogers there, and then continue north.

The pontoons that Post ordered never arrived so, instead, he had a set installed that were meant for a much larger plane. They made the hybrid aircraft, which was already nose-heavy, even more so.

Rogers commented on the huge pontoons, but Post reassured him that they would be fine, and a test flight from Seattle in early July 1935 seemed to confirm that the plane was airworthy. When Rogers and Post decided that they would add hunting and fishing to their Alaskan itinerary, Mae decided to leave the two men to their own devices and head for home. On 6 August, the Orion-Explorer took off, further laden down with the expedition's gear and two cases of chilli con carne, arriving in Juneau, Alaska, the next day. There, Post and Rogers visited friends and, owing to poor weather, stayed a few days before flying on to Dawson in the Yukon Territory, then on to Fairbanks, with Post doing the piloting while Rogers tapped out his columns on his typewriter.

On 15 August, the pair left Fairbanks for Point Barrow, where Rogers hoped to interview an elderly whaler for a story he was working on. Having refuelled at Lake Harding, they were getting close to their destination when bad weather closed in and, uncertain of their position, they decided to land in a lagoon to ask directions. They chatted for a while with Clair Okpeaha, the proprietor of a nearby sealing camp, had some food, and then climbed back into the aircraft for the final leg. Barely had the seaplane lifted off when the engine stalled. With no power to keep the nose up, the front-heavy plane plunged into the lagoon, rupturing its fuselage and ripping off its right wing, finishing upside down in the shallow water. Both men died on the spot.

Okpeaha immediately hurried sixteen miles, on foot, to Barrow to get help. Hours later, a rescue party located the bodies, noting that Post's watch had stopped at 8.18 p.m. on 15 August.

But his legacy lived on. Astronaut Thomas Stafford said, 'Every time I donned a modern spaccsuit, I thought of Wiley Post.' Amelia Earhart, one of the greatest of female fliers, described him as 'the most courageous pilot in the history of aviation'.

# 6

## FLYING IN THE FACE OF REASON

Mention the name Howard Hughes and the first two words that might spring to mind are 'wealthy' and 'weird'. But although he was both of these (and increasingly so with age), as well as a movie producer, sportsman, legendary womanizer, and Las Vegas mogul, his greatest achievements were in aviation. He built and brilliantly flew some extraordinary planes, personally smashed virtually every airspeed record in his day, and was a major architect of the American airline industry.

Hughes was probably born in Houston, Texas, although according to some accounts it was in the little town of Humble. His birth date may have been Christmas Eve 1905 – the day he himself quoted – but his baptismal record begged to differ, listing it as 24 September. Right from the start, then, Howard Hughes was an enigma, a difficult character to pin down or understand.

One thing that is certain is that his wealth came from his dad, Howard Sr, a wildcat oilman who patented a rotary drill-bit that revolutionized oil drilling worldwide. Through their firm, the Hughes Tool Company, Howard Sr and his partner cannily leased rather than sold the bits for as much as $30,000 per well, amassing personal fortunes in the process.

Howard Jr, meantime, while not a great achiever in school, showed an early flair for engineering. He built the first radio transmitter in Houston at the age of eleven, then put together a motorized bicycle using parts taken from a steam engine of his father's. At fourteen he started flying lessons. Two years later his mother died, then two years after that his father passed away as well, leaving Hughes to inherit the bulk of the family's riches plus a large ongoing income from the tool company.

## MR HUGHES GOES TO HOLLYWOOD

Although reclusive by nature and not very personable or articulate, the tall twenty-year-old millionaire had an eye for the glamorous life and, shortly after his father's death, moved to Los Angeles with his new wife to try his hand at movie making. His first two films were commercial successes, and one of them won an Academy Award. Then in 1930 he took on an epic movie that even Steven Spielberg might have quailed at. Called *Hell's Angels*, it was about Royal Air Force fighter pilots in the First World War and cost $3.8 million to produce. Much of this huge outlay (equivalent to maybe $100 million or more today) was spent on amassing the largest private air force in the world – eighty-seven vintage Spads, Fokkers and Sopwith Camels – and housing and maintaining it. Three stunt pilots died while filming the aerial combat scenes, which Hughes personally directed, and Hughes himself almost came to grief when his scout plane crashed and he had to be hauled out of the wreckage unconscious, his cheekbone crushed.

*Hell's Angels* was a runaway box-office hit and secured Hughes's status as a young Hollywood legend. His wife, less than impressed by his affairs with young actresses, headed back to Texas and divorced him. But the fact is that, although Hughes was often in the company of alluring women such as Ava Gardner, Ginger Rogers, Lana Turner, and Katharine Hepburn, not to mention a few leading men, his greatest passion, from a very early age, was for aeroplanes.

In 1934 he won his first race and set a national speed record in a highly modified Boeing 100A biplane, averaging 185 miles per hour. That same year he brought together a small team of engineers to work on a project that would go down in aviation history. The result was one of the most beautiful planes ever built, and certainly the fastest of its time – the H-1, or simply 'the Racer' as Hughes preferred to call it. Essentially a technology demonstrator, it featured a number of design innovations: retractable landing gear, flush rivets and joints to reduce drag, and a monocoque stressed skin. Driven by a muscular 1,000-horsepower Pratt & Whitney Twin Wasp engine, it powered Hughes to a new airspeed record of 352 miles per hour (566 kilometres per hour) on 13 September 1935, over a test course near Santa Ana, California, before landing ungraciously, its gear still up, in a bean field.

A year and a half later, in a redesigned Racer with longer wings, Hughes set a new transcontinental record by flying non-stop from Los Angeles to Newark in seven hours twenty-eight minutes, knocking two hours off his own previous best time for the run. For much of the way he had flown above 15,000 feet

**14** Howard Hughes standing in front of his new Boeing Army Pursuit Plane (Boeing 100A) in Inglewood, California in the 1940s.

without any special oxygen equipment – a reckless act in the eyes of some aviation professionals. The remarkable, unique H-1 never flew again and was retired after a mere forty-two hours of flight time. But it had whetted Hughes's appetite to take on more records.

## LIFE IN THE FAST LANE

The ambitious aviator was determined to be the fastest man around the planet. For his round-the-world record attempt he bought the only Douglas DC-1 ever built, but in the end chose to use a Lockheed 14 Super Electra, equipped with two enormous Wright Cyclone engines (the most powerful available), extra fuel tanks, and navigation equipment. On 10 July 1938, with a crew of three, Hughes set off from Floyd Bennett Field in New York, landing in Paris (breaking the New York to Paris record previously held by Charles Lindbergh), Omsk, Yakutsk, Fairbanks, and Minneapolis, before arriving back at the starting point just ninety-one hours later, having shaved four hours off the previous best time.

That wasn't enough for Hughes. He wanted to smash the record again, in a bigger plane and in more style, to show everyone that the age of safe, comfortable, long-distance air travel had arrived. So he bought a Boeing 307 Stratoliner – the world's first pressurized airliner – and had the usual long-range modifications made to it, including extra fuel tanks. But the start of the Second World War made air travel across Europe unsafe and forced him to revise his plans. Later, he had the Stratoliner kitted out with a deluxe interior, including a bar, kitchen, powder room and sleeping quarters, with some of the décor suggestions coming from Rita Hayworth, but he never really liked the plane and eventually sold it.

Alongside his record-breaking attempts, Hughes was getting more and more involved with the business side of aviation. The small team he had assembled to build the H-1 Racer went on to form the Hughes Aircraft Company, which first operated out of a rented hangar in Burbank. In 1939 Hughes became a major player in the commercial airline industry when he started buying up stock in Trans World Airlines (TWA). Two years later he became a majority shareholder and took control of the company.

Hughes had been sketching out a concept for a new airliner that would be bigger, faster, and more elegant than anything that had come before. But being now the owner of an airline company meant that, by federal law, he couldn't manufacture passenger planes of his own. So he approached Boeing's main competitor, Lockheed, with the idea of building an aircraft to replace TWA's fleet of Stratoliners. Hughes's good relationship with Lockheed stemmed from the time they supplied the aircraft he used in his 1938 record round-the-world flight. Lockheed agreed to build the new aircraft in secrecy, and the result was the revolutionary Constellation, of which TWA bought forty, straight off the production line.

With its four stylishly cowled engines, distinctive triple tail, and dolphin-shaped fuselage, the Lockheed Constellation became a design icon. The most refined airliner of its day, beloved by pilots and passengers alike, it was ideally placed to take advantage of the postwar boom in intercontinental travel.

But meanwhile there was a world war to be fought, and when the US became embroiled in it TWA temporarily handed over the rights to its embryonic Constellation fleet to the United

States Army Air Force (USAAF), which designated the craft the C-69 and used it as a military transport. In 1944, Hughes and TWA president Jack Frye personally delivered the first Constellation to USAAF, sharing a new coast-to-coast record of just under seven hours.

## NEAR-DEATH EXPERIENCE

At Hughes Aircraft – Howard's other big aviation concern – staff were also gearing up for the war effort. The Hughes D-2 was conceived in 1939 as a fighter-bomber but morphed over time into a two-man reconnaissance aircraft, designated the D-2A, the prototype of which was flown for the first time from Harper Dry Lake, California, on 20 June 1943. In November

**15** The first prototype of the Hughes XF-11, *c*.1946.

the following year, the hangar housing the D-2A was struck by lightning and the aircraft was destroyed. But development continued of another spy plane offspring of the D-2, known as the Hughes XF-11. Able to fly at 400 miles per hour and at high altitudes, it was intended to evade enemy radar and take covert pictures using a newly devised fine-grain film. Its twin twenty-eight-cylinder, turbo-supercharged engines powered counter-rotating double propellers designed to provide extra thrust. Its only test pilot was Hughes himself.

On 7 July 1946, thirty minutes into a flight over Los Angeles, the XF-11 hit trouble. An oil leak in a gearbox caused one of the contra-rotating propellers to reverse pitch, which in turn yanked the aircraft's nose to one side. Hughes tried to save the craft and himself by heading for the only makeshift strip available – the Los Angeles Country Club golf course – but, seconds short of its impromptu target, the XF-11 dropped steeply and came down in a residential part of Beverley Hills. It demolished three houses and its fuel tanks exploded before finally it came to rest. Somehow Hughes managed to drag himself free of the flaming wreckage and collapse nearby. He was then pulled to safety by a sergeant in the US Marines, who happened to be visiting friends in the area. In hospital, the extent of Hughes's injuries became clear. As well as numerous third-degree burns, he had a crushed collarbone, multiple cracked ribs, and such a badly caved-in chest that his left lung had collapsed and his heart had been pushed to the right side of his chest cavity. Although he recuperated, he was left in pain for the rest of his life – almost certainly an important factor in his later addiction to codeine, which he used to inject into his muscles. He also took to wearing a moustache to cover the scarring from the accident on his upper lip.

# Hughes H-4 Hercules "Spruce Goose"

**Crew:** 3
**Length:** 218 ft 8 in (66.65m)
**Wingspan:** 320 ft 11 in (97.54m)
**Loaded weight:** 400,000 lbs (180,000 kgs)
**Powerplant:** 8 x Pratt & Whitney radial engines (3,000 hp each)
**Cruise speed:** 250 mph (407 km/h)

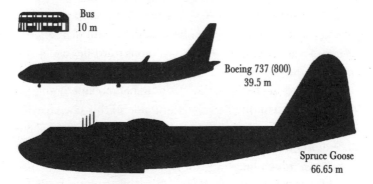

Bus
10 m

Boeing 737 (800)
39.5 m

Spruce Goose
66.65 m

**16** The H-4 Hercules, better known as the 'Spruce Goose'.

A second XF-11 was built, this time equipped with conventional four-blade props, and flown successfully by Hughes, but the project was scrapped shortly after being handed over to the Army Air Force in 1949. It was one of two aircraft for which Hughes had won a wartime contract from the US government. The other was for an astonishing plane, as eccentric in its way as its creator, and that has become indelibly linked with Hughes in the public mind.

## FLYING GEESE AND WHITE ELEPHANTS

The vessel for which Hughes is best remembered was an outrageously large flying boat designed to carry up to sixty tons of cargo or 750 troops across the Atlantic as an alternative to seagoing vessels, which were vulnerable to attack by German U-boats. Known originally as the HK-1 and later redesignated the H-4 Hercules, it was built, as stipulated in the contract, not from aluminium, which was considered a critical resource during the war, but from wood. This led to it being popularly called the 'Spruce Goose', because it was a large aquatic flier and 'spruce' rhymes with 'goose' – although the wood actually used was birch. Less flatteringly, it was also known as the 'Flying Lumberyard'.

Although the contract for the giant plane was signed in 1942, due to Hughes's procrastination and obsession with detail the H-4 was still unfinished at the end of the war. Only on 2 November 1947, long after there was any real use for it or logic for its existence, was the Spruce Goose taxied out into Long Beach harbour in California for its first and only flight. It was an awesome sight – the largest plane ever built up to that time (twenty per cent bigger than a 747), with wings, each supporting four massive engines, spanning 97.5 metres (still, today, the longest wingspan of any aircraft in history) and a tail that soared seven storeys high.

At the controls was – who else? – Hughes himself, supremely confident that the 180-ton bulk of his $25 million behemoth would rise into the air. 'If it fails to fly,' he had said earlier, 'I will leave the country. And I mean it.' With a gaggle of reporters and a small curious crowd watching from the shore (the test hadn't been widely advertised), Hughes throttled up the eight 4,000-horsepower Pratt & Whitney engines. Slowly, the Goose gathered speed, kicking up spray and smacking against the low waves as it strained to get airborne. To those watching, it must have seemed that the great seaplane was in danger of running out of harbour. But finally it rose, all of sixty feet out of the water, and travelled a mile before flopping back down. The only journey of the magnificent, but now completely pointless, Spruce Goose had lasted less than a minute.

With the best will in the world, even if the US government had felt motivated to throw millions more dollars at the project in peacetime (which it didn't), it is hard to see how Howard's folly could ever have served a useful purpose. With the throttle wide open, not much fuel in its tanks, a crew and passenger complement of just thirty-two, and most of the cavernous space inside the fuselage empty, the Goose had struggled mightily to achieve flight. Whether it could have reached its service ceiling of around 20,000 feet and intended cruise speed of more than 250 miles per hour, fully laden, will never be known. After its sole excursion, the plane was moved to a dry-dock, cocooned inside a huge, climate-controlled hangar, and never seen by the public again for thirty years.

## DESCENT INTO OBLIVION

The latter part of Hughes's life makes for depressing reading. It was marked by a gradual descent into worsening mental illness and increasingly bizarre behaviour. As early as the 1930s he had shown signs of obsessive-compulsive disorder, becoming fixated on trivial details, such as arranging peas (one of his favourite foods) on his plate by size using a special fork. His reclusiveness became extreme. In 1957, Hughes shut himself away for four months in the screening room of a film studio near his home, surviving exclusively on chicken, chocolate, and milk, and not leaving even to wash or go to the bathroom. He sent memos to his aides telling them to speak only when spoken to and not to look at him. When he finally came back into the light of day, he was in an appalling state.

Although Hughes was still lucid at times and managed to attend more or less to his various businesses, he became increasingly isolated from the outside world, alternating between almost complete self-neglect – his hair and nails left uncut for a year at a time – and an irrational terror of being exposed to germs. In 1966 he moved into the Desert Inn Hotel in Las Vegas. When asked to leave because of his unacceptable habits, he proceeded to buy the place rather than be evicted, along with four other casinos in the town, a radio station, and other Nevada properties.

Hughes now lived in perpetual agony from his past injuries, to the extent that he avoided even brushing his teeth because the action was too painful. A doctor who examined him in 1973 likened his physical condition to that of POWs he had seen in Japanese prison camps during the Second World War. Towards the end of his life, the billionaire eccentric moved to the Bahamas and then to Mexico so that, reportedly, he could have greater access to the codeine with which he regularly injected himself. He died on 5 April 1976 of apparent heart failure, perhaps fittingly while aboard a plane en route from Acapulco to a hospital in his hometown of Houston. One of the giants of the aviation world, and an aerial speed pioneer, had passed away not having been seen in public or even photographed for twenty years.

# 7

## JOHN STAPP AND HIS INCREDIBLE SLEDS

US Air Force flight surgeon John Paul Stapp was strapped, forward-facing, into an exposed chair on the back of a rocket sled that could travel faster than a Jumbo Jet. Ahead of him was a 2,000-foot-long rail track stretching across the arid landscape of Muroc Army Airfield in California. Stapp himself counted down, his voice sounding over the intercom in the control room some distance away. At 'zero', the rockets were ignited and the sled accelerated ferociously over the next five seconds to an astonishing 632 miles per hour, demolishing the land-speed record.

Moments later, the sled ploughed into a trough of water, which brought it to a brutally abrupt halt. No human before or since has willingly undergone such a jolt: from over 600 miles per hour to rest in a little over a second – the equivalent of hitting a brick wall at 120 miles per hour. So fast had Stapp travelled that dust particles had speared through his flight suit, raising blisters on his body. So suddenly had he slowed down that the capillaries in his eyes burst, his eyeballs bulged from their sockets, and he was left temporarily blinded. The date was 10 December 1954.

What may have seemed like an act of mindless masochism was just one of many extreme experiments for which Stapp volunteered in the name of science. His goal: to learn how humans could best survive the trauma of crashing aircraft.

## INSPIRATIONS

Born in Bahia, northern Brazil, in 1910, the son of Baptist missionaries, Stapp moved with his family to the United States at the age of thirteen, and later started college in Texas with the idea of becoming a writer. During a Christmas break in 1928, he witnessed a tragedy that changed the course of his life. While he was visiting relatives, his baby cousin crawled into a fireplace and was badly burned. The child died three days later, but for those three days Stapp helped nurse the infant and afterwards determined to become a doctor. Fifteen years later, having earned degrees in zoology and biophysics, he entered the University of Minnesota Medical School to pursue his dream. Upon graduation, he interned at St Mary's Hospital, Duluth, before enlisting in the Army Medical Corps towards the end of the Second World War.

In 1946, Stapp was transferred to the Aero Medical Laboratory at Wright Field (now Wright-Patterson Air Force Base), in Ohio, as a project officer and medical consultant in the biophysics branch. His first assignment included a series of flights in an unpressurized aircraft at heights of 40,000 feet or more. This was a time when newly developed military jets were flying faster and higher than ever before, increasing the risk of decompression sickness (DCS), or 'the bends' – an excruciatingly painful and potentially fatal condition, first noted in aviation in the 1930s during flights of high-altitude balloons and high-flying propeller-driven aircraft. Under normal conditions, some gases from the atmosphere – mainly nitrogen – are dissolved in a person's blood. If that person then goes quickly to a much higher altitude (or place of lower pressure), the gases can come out of solution in the blood and form bubbles. As the bubbles try to escape, they give rise to a variety of symptoms, the most common of which is mild to extreme pain in the joints, especially the shoulders, elbows, knees and ankles. In the worst cases, the individual twists and doubles up in agony – hence 'the bends' – an instinctive reaction that only makes the condition worse.

Before the days of high-flying balloons and aircraft, only divers and miners tended to suffer from DCS, which happened on rare occasions if they came up too quickly from depths below the surface of the ocean or earth. The condition could be avoided by making a gradual ascent – in the case of divers about ten metres (thirty-three feet) per minute. But such a leisurely rate of ascent and descent is not acceptable in flight, and the only practical alternatives are to adjust the pressurization of gas in the aircraft cabin, or to wear individual pressure suits supplied with oxygen at a pressure that automatically adjusts according to the altitude. It was to study how people react to loss of pressure at high altitude, and how best to address the problem, that experiments had to be done on human guinea pigs in the immediate postwar years.

To carry out these tests, a Boeing B-17 Flying Fortress – a four-engined, prop-driven heavy bomber much used against German targets – was drastically modified to be able to fly into the stratosphere. Stripped of all weaponry and with redesigned engines, it could cruise for hours at altitudes of nearly 45,000 feet. In the spring of 1946, this war-machine-turned-research-vehicle took to the air, its interior unpressurized and unheated, with a flight crew condemned to shiver in sub-zero temperatures, while at the back, alone, sat Captain Stapp, following one of the unwritten rules of human volunteer testing – whatever you would like to subject others to, you have to be willing to visit on yourself.

## AT THE EDGE OF SURVIVAL

The questions Stapp was trying to answer were critical to the future of aviation. How would the body react to prolonged exposure at high altitude? Could humans properly function, physically and mentally, under this regime? How could they keep themselves from freezing, dehydrating, or becoming incapacitated by the bends? After a total of sixty-five hours, flying half as high again as Mount Everest, in conditions even Ryanair passengers wouldn't tolerate, Stapp had accumulated a mass of invaluable data. Thanks largely to his efforts, physiological issues were resolved that smoothed the way for the next generation of high-altitude aircraft. They also helped refine techniques for what became known as HALO (High Altitude Low Opening) jumps, which eventually involved individuals leaping from balloon-borne capsules at the edge of space. Avoiding the bends at high altitude proved a particularly tough nut to crack. In the end, Stapp's experiments showed that breathing pure oxygen for half an hour prior to take-off would eliminate the symptoms entirely. This was a crucial breakthrough in the development of systems to avoid depressurization effects.

As a dubious reward for his pioneering work on high-altitude flight, Stapp was assigned to head up the Aero Med Lab's top-priority project, code-named MX981: the study of severe deceleration on the human body. It had become a rule of thumb in the military that the most extreme jolt that a person could survive during a collision was 18g (where 1g is the force of gravity acting on a body at sea level). As a result, aeroplane cockpits were not built to withstand anything greater than this g-force. Yet during the Second World War evidence emerged that cast serious doubt on the 18g threshold. There were cases where naval pilots had slammed into the islands of aircraft carriers, or into other aircraft, so hard and fast that, according to accepted wisdom, they should have died instantly. Instead they walked away with barely a scratch. More troubling were the many, less violent crashes in which it seemed clear that the pilot had survived the initial impact only to die due to the failure of the seat, harness, or cockpit structure.

These incidents underscored the need for rigorous scientific studies of how air safety could be improved. What were the true levels of g-force that the human body could tolerate, and for how long? What was the maximum speed at which a pilot could safely eject? How could seats, harnesses, and other equipment be developed to better protect their users?

## GEE WIZARD

In April 1947, Stapp travelled to Los Angeles to check out the 'human decelerator' being built at Muroc Army Airfield. A key element of this was a 2,000-feet-long track laid down during the war for testing German V-1 missiles (the infamous 'buzz

**17** John Stapp rides the Gee Wizard at Muroc Army Airfield.

bombs'). At one end of the track, engineers from Northrop Corporation had fitted sets of hydraulic brakes capable of slowing a rocket sled from 150 miles per hour to half that speed in one-fifth of a second – a deceleration equivalent to that experienced in a plane crash.

The sled itself, nicknamed Gee Whiz, was built of welded tubes, measured fourteen feet long and just over six feet wide, weighed about 1,350 pounds, and sat on a series of magnesium slippers. On top of the chassis was a tough, specially built seat, and to the rear a telemetry antenna mast and a rack that could hold up to four rocket bottles. These bottles, the same type used to boost aircraft off short runways, could each produce about 5,000 pounds of thrust. By varying the number of bottles and the brake pressure, the occupant could be subjected to a wide variety of unpleasant g-forces.

The occupant was intended to have been a 185-pound dummy known as Oscar Eightball. The staff at the Aero Med Lab hadn't even contemplated running human tests because of the obvious risks. However, all that changed with the arrival of Stapp. Having introduced himself to George Nichols, Northrop's top engineer on the site, Stapp walked over to Oscar Eightball, patted it on the head, and said: 'You can throw this away. I'm going to be the test subject.'

But Stapp was no reckless fool. Oscar would still make the first trips on the Whiz – and a good thing, too. During the initial run, on 30 April 1947, the hydraulic brakes and back-up restraint system failed, and the Whiz slid off the track into the desert. In a later test, in which the intrepid dummy was blasted down the track at 150 miles per hour wearing only a light safety belt, the brakes locked producing a savage deceleration, the belt snapped, and Oscar sailed through an inch-thick wooden windscreen, leaving his rubber face behind, before finally coming to rest in a mangled heap 700 feet down range.

After eight months and thirty-five test runs using the hapless manikin, Stapp was ready to try out the Whiz himself. For his first outing he used only one rocket and sat facing backwards to minimize the effects of the g-load. Suffering only a few sore muscles from the experience, the next day he added two more rockets, and so the experimentation went on. With each new run, Stapp tinkered with the thrust and braking not only to step up the g-forces but also the time it took for the forces to build to maximum (the so-called rate of onset) and their duration. By August 1948, Stapp had made sixteen runs and endured up to 35g – far beyond the supposedly lethal 18g – without any survival limit in sight. Not that Stapp escaped the process without discomfort or injury. Every time he rode the Whiz, his seat harness bit painfully into his shoulders, and at higher g's his ribs occasionally cracked under the pressure. He also suffered concussions, lost fillings from his teeth, fractured his collarbone, and broke his wrist (twice) when his arm flapped free in the vicious airstream. On one occasion, he set the fracture himself on the way back to his office. But of all the medical issues Stapp encountered, the most worrying centred on his eyes.

During backwards-facing tests the big problem was 'white outs' caused by the blood suddenly draining from his eyeballs and pooling at the back of his head. In later runs, when he faced forwards, the blood rushed the other way and pushed against his retinas, rupturing capillaries, filling the front of his eyes with blood (a horrific sight to onlookers), and causing 'red outs'.

After many months, Stapp's superiors at the Aero Med Labs got wise to what he was up to and were appalled. Stapp was promoted to major (a move designed to keep him busy with paperwork), and was given a reminder of the officially recognized 18g survivability limit, and an admonition to stop deceleration tests on people (notably himself). From now on, the brass insisted, if there were any live experimentations, they would be done on chimpanzees.

But that decision was soon reversed. As the import of Stapp's findings sank in, it became obvious that human tests had been key to exposing the inadequacy of certain types of aircraft restraint systems. New designs were brought in almost right away. The advantage of rear-facing seats in a collision was also driven home by the new data, and from that point on all new transport aircraft were fitted with this kind of seating. Most significantly, Stapp had put the lie to the 18g limit. The fact that the human body could withstand forces of 30g or more, for brief periods, meant that seats, harnesses, and cockpits had to be beefed up to remain secure beyond that limit.

The next series of rocket-sled tests, using a new heavy-weight harness that allowed forward-facing runs, was aimed at refining the data. Beginning in June 1949, the Whiz was pushed to further extremes, sometimes with Stapp aboard, sometimes with other human volunteers, and sometimes with chimps. And it was around this time that Murphy's Law was born. There really was a Murphy – an air-force engineer called Captain Ed Murphy who came out to Muroc with some new strain gauges to give a more precise measure of the decelerations being produced by the Whiz. One of Murphy's assistants fitted the gauges to Stapp's harness but managed inadvertently to wire up two of them the wrong way, with the result that all the readings cancelled out. On this particular run, Stapp had a rough ride, and ended up staggering away from the sled with bloodshot eyes and bleeding sores. Due to the miswired sensors, however, the gauges read zero, literally nullifying Stapp's efforts – and Murphy was not pleased. 'If there's any way they can do it wrong,' he remarked, 'they will.' His comment made the rounds at the track and ended up gaining worldwide fame when paraphrased by Stapp during a press conference.

In June 1951, Stapp made his last run on the Whiz, soaking up more than 35g – about as much punishment as the sled could mete out with all four rockets blazing and the brakes at their peak setting. In total, it had been used for seventy-four manned runs and eighty others with chimps. The tests established a standard strength requirement for aircraft seats of 32g, which was rapidly put into practice.

## THE FASTEST MAN ON EARTH

Yet while the Whiz had helped answer many questions about crash deceleration, a new conundrum had emerged. In 1951, no one had yet ejected from an aircraft at supersonic speed and lived to tell the tale. Very little was known about the effects of wind blast and deceleration acting on a pilot bailing out in that regime. But, sooner or later, pilots would be forced into attempting supersonic escapes. Would they survive, and what could be done to improve their chances?

The Gee Whiz had reached its limits. So, in 1953, Stapp relocated to the Aeromedical Field Laboratory at Holloman Air Force Base in New Mexico, where a much longer sled track was in place, terminating in a segment that could be dammed and filled with water. By equipping a sled with water scoops, and varying the water depth precisely, various braking speeds and durations could be produced.

Northrop built a new sled, called the Sonic Wind No. 1, which was slightly longer and wider than the Whiz, and could carry up to twelve rockets with a combined thrust of 50,000 pounds. It also had a two-stage design: after the rocket bottles burned out, the 'propulsion sled' was jettisoned, allowing the 'subject sled' to continue on without the extra load. In theory, the Sonic Wind No. 1 could travel up to 750 miles per hour, and deliver a bone-crushing 150g.

**18** Upper: Stapp is prepared for his record-breaking run aboard Sonic Wind No. 1. Lower: Sonic Wind No. 1 hits the water trough that slowed it from 632 miles per hour to rest in little over a second.

In November 1953 the Sonic Wind No. 1 was put through its paces with Sierra Sam, an upgraded test dummy. A few months later, Stapp made his first trip down the new track. 'I assure you,' he quipped to a reporter as he boarded, 'I'm not looking forward to this.' Burning six rockets, the Sonic Wind No. 1 topped out at 421 miles per hour after just five seconds, and was still doing 313 miles per hour when it hit the water brake. In the space of 190 feet, the sled slowed to 153 miles per hour, producing up to 22g. For a few moments Stapp felt as if he weighed 3,500 pounds, and for over half a second he endured 15g. That was nearly twice as long as any similar effect ever produced at Muroc. 'I feel fine,' Stapp said after the run, 'this sled is going to be a wonderful test instrument.'

So it proved. On 10 December 1954, after many more gruelling but fruitful sessions on the Sonic Wind No. 1, Stapp took the sled chair for his twenty-ninth, final, and most extreme ride. He would push into the transonic zone – near the speed of sound – reaching Mach 0.9. He would be facing into the onrushing airstream, protected only by a helmet, visor, and regulation flight suit. His major concern was that he would lose his sight. Prior to this last trip he had 'practised dressing and undressing with the lights out so if I was blinded I wouldn't be helpless'.

With ten minutes to go before the big run, George Nichols helped fit a rubber bite block, equipped with an accelerometer, into Stapp's mouth. As the seconds ticked away, a Lockheed T-33 Shooting Star chase plane approached, its throttle pushed wide open in anticipation of the launch. With five seconds to go, Stapp activated the sled's movie cameras, and hunkered down for the mind-numbing jolt. The Sonic Wind No. 1's

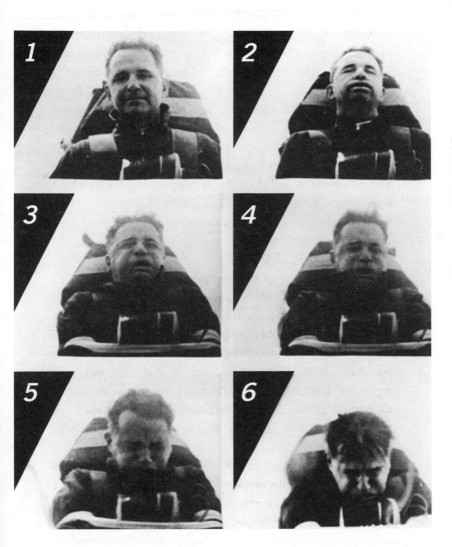

**19** Stapp's face shows the effect of a high-speed trip aboard
Sonic Wind No. 1.

nine rockets roared into life, spewing thirty-foot flames, and hurling Stapp down the track and past the chase plane as if it were standing still. A few seconds into the run, the sled had peaked out at 632 miles per hour, subjecting Stapp to 20g and battering him with wind pressures of nearly two tons. Then, just as the sound of the rockets' initial blast reached the ears of far-off observers, the Sonic Wind No. 1 hit the water brake, throwing out spray thirty yards on either side, before grinding to a halt as if it had hit a sandbank. Stapp had maxed out at 40g deceleration, and temporarily blacked out. The ground crew raced to the scene, followed by an ambulance. An agitated Nichols leapt onto the sled, anxious to assure himself that Stapp was still alive.

His eyes were a mess, filled with blood. When Sonic Wind No. 1 had hit the water brake, it was the equivalent of a Mach 1.6 ejection at 40,000 feet, only it had lasted perhaps nine times longer. As attendants helped the injured man onto a waiting stretcher, even Stapp worried that he had pushed his luck too far. 'This time,' he remarked, 'I get the white cane and the seeing-eye dog.' But when surgeons at the hospital examined him, they found that his retinas hadn't detached as feared. Within minutes, Stapp could see 'blue specks' and a short time later could discern one of the surgeons' fingers. By the next day, his vision had returned more or less to normal.

But Stapp's life would never be the same again. Dubbed 'The Fastest Man on Earth' by the media, he became a scientific celebrity. He appeared on the pages of *Collier's* and *Life* magazines, was the subject of a Hollywood 'B' movie, and featured in an episode of *This is Your Life*.

Meanwhile, he announced plans to crack the sound barrier

on land, and, beyond that, attempt a 1,000-miles-per-hour run. But those were steps too far for his superiors – and their caution turned out to be justified. In June 1956, while engaged in an 80g test, Sonic Wind No. 1 left its track and was badly damaged. Human experimentation was put on hold, and although Stapp would take part in subsequent tests on an air-powered sled known as the Daisy Track, his days as a rocket man were over.

# 8

# THE X-MEN

America's first civilian astronaut never flew aboard a space-craft. On 19 July 1963, Joe Walker took his rocket-powered aircraft above a height of sixty-two miles (one hundred kilometres) – the so-called Kármán line – which by some definitions marks the boundary between Earth's atmosphere and outer space. The plane was the North American X-15, and Walker was among a select group of individuals to fly aboard one of the extraordinary X-series of planes. Another X-plane pilot was the first human to travel faster than sound, Chuck Yeager.

The first X-plane (X for 'experimental') was conceived in 1944, at the tail end of the Second World War, by the US Army Air Force and NACA (National Advisory Committee for Aeronautics) – the forerunner of NASA. Its goal was straight-forward: to break the sound barrier. Bell Aircraft Corporation was given the contract to build the plane, and at the end of 1945 handed over the first Bell X-1 for testing.

The X-1 was designed to travel as fast as a proverbial 'speeding bullet', so it made sense that it should be shaped like a bullet with wings. The fuselage was modelled after a Browning .50-calibre machine-gun bullet, which was known

**20** The Bell Aircraft Corporation X-1 with shock-wave pattern
visible in its exhaust plume.

to be stable in supersonic flight. No bulges – such as a bubble
cockpit – were allowed, so the pilot had very poor visibility. In
fact, from where he sat, he had no view at all to the rear and
only a very poor view looking forward, given the gentle slope
to the nose and the bank of dials and switches that were piled
high in front of his face.

There was also a problem of how to escape from the plane
if something went wrong as the bullet shape didn't provide
enough room for an ejection seat. There was a little side hatch
that the pilot could squeeze out of in a dire emergency. But
he would have to be pretty desperate, because right behind
the hatch was the wing, and if by some miracle he missed
that then he'd likely be thrown up into the tail. Given that the
choice was between staying with the plane or ending up as
mincemeat, it's no surprise to hear that the X-1 programme
didn't see a single bail-out.

As far as propulsion went, the X-1 was all rocket. The
rocket motor had four chambers and there was a choice to
fire one or more chambers at a time, but once they had been

ignited that was it: they would burn full on until all the liquid fuel was used up or they were shut off completely (rockets that could be throttled came later). At maximum thrust, with all four chambers firing, the X-1 had just two and a half minutes of power. To save its precious fuel for boosting it to record-high speeds, rather than the mundane tasks of getting off the ground and climbing the first twenty-odd thousand feet, the X-1 was carried aloft by a B-29 Superfortress and mated to its mothership by a standard heavy-duty bomb shackle.

## ONWARDS AND UPWARDS

The name Charles 'Chuck' Yeager is synonymous with the Bell X-1 and the smashing of the sound barrier. But he wasn't the first to take the controls of the flying bullet. That honour went to Bell Aircraft's chief test pilot, Jack Woolams. In September 1942, although only in his late twenties, Woolams had already become the first person to fly a fighter non-stop from coast to coast across the United States and, the following summer, had cracked the world altitude record for an aircraft, reaching 47,600 feet.

Woolams was a consummate prankster. During the war he was at the Materiel Command Flight Test Base (later part of Edwards Air Force Base), California, putting the new Bell XP-59A through its paces. This was America's first military jet and a hush-hush project, so that when it was towed from its hangar to the run-up area, a fake propeller was attached to the nose to disguise its true nature. Still, Woolams couldn't resist the opportunity for a practical joke, even if it meant breaching security. During a daylight XP-59A mission in autumn 1943, he noticed a Lockheed P-38 Lightning from a nearby training unit flying in the same area. Taking off his flight helmet, he slipped on a furry mask and short-brimmed hat, and eased the jet alongside the P-38. The Lightning's pilot must have got the shock of his life when he saw the sleek, propeller-less plane, barely twenty feet away, being flown by what appeared to be a gorilla in a derby hat – waving a cigar.

In January 1946 Woolams found himself in a very different situation. He was 8,000 feet up in a B-29, climbing down a ladder and into the tiny cockpit of the Bell X-1 nestled partly inside the bomber's dark underbelly. At 28,000 feet the count-down to launch began and, at zero, the small rocket plane was released into the dazzling light of day for its maiden flight – an unpowered cruise back to Pinecastle Army Airfield near Orlando, Florida. Nine more such outings followed before the first X-1 was taken back to Bell Aircraft for some modifications. A new series of tests was to begin in September 1946, but they would be flown by another pilot.

At the end of August, Woolams was flying over Lake Ontario in a Bell P-39 Airacobra – a single-seater that had been one of the most successful American fighter air-craft of the Second World War. He was preparing for the upcoming National Air Races in Cleveland, and nudging the P-39 up towards 400 miles per hour. Suddenly and inexplicably it went out of control, plunging into the water and breaking up on impact. Woolams's body was recovered four days later.

The Bell X-1 tests resumed, but by the following summer the air force had grown impatient with the progress of the project. Bell Aircraft were running what the brass considered to be too slow a build-up to the assault on Mach 1 – the speed

**21** Chuck Yeager standing alongside the Bell X-1, which he nicknamed 'Glamorous Glennis' after his wife.

of sound. Bell's contract for testing was cancelled and responsibility for going through the sound barrier was handed to the Army Air Force Flight Test Division (the 'Army' part of that title was soon to be dropped as the US Air Force became a separate service in September 1947). The new chief test pilot was Charles 'Chuck' Yeager, a Second World War veteran blessed with astonishingly acute 20/10 vision (which once enabled him to shoot a deer at 600 yards). During the war, he was the first pilot in his P-51 Mustang group to 'ace in a day', accounting for five enemy planes in a single mission. He was also one of the few Allied pilots to shoot down a German jet, a Messerschmitt Me 262.

## THROUGH THE BARRIER

After a couple of months flying the X-1, Yeager was nudging it very close to the sound barrier. At speeds of Mach 0.95 – ninety-five per cent of the speed of sound – the plane was getting buffeted a lot by the turbulent air piling up around it. No one quite knew what was going to happen when the aircraft finally went supersonic, because there was no useful wind-tunnel data. Models had been put in wind-tunnels and subjected to air moving at supersonic speeds but the shock that formed on the models at about Mach 0.9 would simply bounce off the walls and block the air flow. So what happened between about ten per cent below the speed of sound and ten per cent above it was more or less a mystery.

The term 'sound barrier' first came into use during the Second World War. Fighter pilots who made high-speed dives noticed several irregularities as flying speeds approached the speed of sound: aerodynamic drag increased markedly, much more than normally associated with increased speed, while lift and manoeuvrability decreased in a similarly unusual way. Pilots at the time mistakenly thought that these effects meant that supersonic flight was impossible; that somehow planes would never travel faster than the speed of sound.

Tuesday 14 October 1947 was the date earmarked for Yeager's next Bell X-1 flight. Although there was no set goal of trying to go supersonic on that day, a sense of anticipation was growing among everyone involved with the project, given how close the plane had been flirting with the barrier. But on the Saturday before the flight, Yeager and his wife Glennis were out horseback riding when Yeager hit a fence that had been closed across a road; he was thrown off his horse and ended up cracking two ribs. Rather than let the flight surgeon know about his injury and risk being grounded, Yeager and his buddy Jack Ridley decided to try a work-around. Flying the plane wasn't going to be a problem. The snag was that, with his injured right side, Yeager wouldn't be able to close the cockpit door with his right arm. The solution: Ridley sawed off a ten-inch length of broomstick so that Yeager, once in the pilot's seat, would have enough leverage to push the locking mechanism closed.

With the B-29 at 8,000 feet, Yeager climbed down into the 250-miles-per-hour ice-cold slipstream, bending double and climbing painfully into the dark little cockpit. A few minutes later came the drop into the blinding light of day, and then a wild ride into the history books. Yeager fired off all four engine chambers, climbed to 35,000 feet, turned off two of the chambers, and continued to climb to 42,000 feet before levelling off and reigniting the third chamber. With the machmeter showing Mach 0.92 he experienced the usual buffeting; at Mach 0.97 the needle suddenly jumped off the scale (the maximum value marked on it was 1.0). At first, Yeager thought the instrument was faulty and radioed 'it's gone kinda screwy on me'. On the ground, a loud bang was heard – not the X-1 breaking up but breaking through the sound barrier. Suddenly the buffeting stopped and Yeager took the rocket plane up to Mach 1.07 (about 650 miles per hour) before gliding back to base and the congratulations of a select handful of people who knew about the achievement. The X-1 project was classified and news that the sound barrier had been crossed wasn't made public until June the following year.

## DANGER AT MACH 2

With supersonic flight now a reality and fears about the sound barrier blown away, there was a major push to move on to much higher speeds. The research was important to the field of aerospace as a whole and the development of new, high-performance fighter planes in particular.

Even as work began on new X-planes – such as the Bell X-2 rocket plane, with its swept (angled-back) wings; and the incredibly slender, jet-powered Douglas X-3 Stiletto – the original Bell X-1, which eventually reached 1,000 miles per hour, evolved into a number of variants. On 12 December 1953 one of these, the Bell X-1A, was piloted by Yeager to another speed record – Mach 2.4. But on this occasion events threatened to get seriously out of hand and only Yeager's experience managed to save the day.

The X-1A was about seven feet longer than the old X-1 and carried almost twice as much fuel so that it could accelerate for much longer. After three flights, Yeager had already cranked the new plane up to Mach 1.9. The fourth flight started well with a good drop and all four rocket chambers firing, powering the plane on a steep climb through 60,000 feet, 70,000 feet, and on up to 80,000 feet. By now the X-1A was passing Mach 2.3, gaining another thirty miles per hour every second. Suddenly, the plane started to yaw, its nose drifting to one side. Yeager responded by pushing the rudder to try to get the nose back in line, but it had no effect. The yaw got worse, and then the outside wing began to rise. The situation quickly became desperate. The aircraft rolled until it was flying upside down, pitched up. The stress on the cockpit canopy was too great and it split open, exposing Yeager to the cold, thin outside air, and causing his pressure suit to inflate.

The X-1A was rolling ferociously, as if it were the most vicious corkscrew roller-coaster you could imagine, rotating twice around per second, and putting Yeager through a withering 9g. Ground and sky flashed by in dizzy succession but, crucially, Yeager stayed conscious and alert to what was happening with the plane. As the X-1A ran out of fuel it slowed, and the rolling stopped. Yeager saw the sky 'below' him and the horizon going round and round, and realized he was in an inverted, flat spin, pulling negative g's – not ideal, but at least he knew what to do about it. Test pilots are as used to spinning planes as test drivers are to making handbrake turns, and they know exactly how to make them stop: set the aileron (the hinged flap on the trailing edge of each wing) with the spin direction, apply the rudder, then fall out of the spin. It worked. The whole ordeal, from the time that the aircraft started to yaw at 80,000 feet until Yeager popped out of the spin at 25,000 feet, lasted a mere 51 seconds, but contained enough stomach-churning action and danger to last most ordinary mortals a lifetime. Throughout it all Yeager had had to contend with a smashed canopy, exposure to the sub-Arctic cold of high altitude, and a bulked-out, inflated pressure suit. But now he was as good as home. Looking around, he spotted his landing site, Rogers Dry Lake, about fifty miles away and glided on back to base. The X-1A had made its first and only excursion above Mach 2.

## AT THE EDGE OF SPACE

Of all the dozens of other X-planes that have been built –
and continue to be built – only one, the X-15, has achieved a
similar, legendary status to the X-1. It too was a rocket plane,
but instead of peaking out at two and a half times the speed
of sound, it eventually reached more than Mach 6 and flew so
high that, for minutes at a stretch on some missions, it was not
an aircraft but a spacecraft plunging through the near-airless
void fifty or sixty miles above the ground. Thirteen flights

**22** Neil Armstrong next to the X-15.

of the X-15 by eight different pilots met the US Air Force's criterion for a spaceflight and earned their pilots the right to be called astronauts.

The X-15 was a collaborative project between the US Air Force and Navy, and NASA. With an airframe built by North American Aviation and an engine supplied by Reaction Motors, the fifty-foot-long plane was the first to be designed specifically to cope with the unusual demands of hypersonic speeds – those above about Mach 5.5. The wedge-shaped tail surfaces were to give directional stability at speeds where aerofoils of a more conventional shape wouldn't have been effective. The large upper and lower fins, and the downward slant of the stubby wings, were intended to keep the aircraft stable during steep climbs and at high altitudes.

Stability was only one of the big challenges of hypersonic flight. Another was overheating caused by friction with the high-speed air. Designers knew that some parts of the plane, like the nose and wing edges, would reach temperatures above 650 degrees Celsius. So they needed a metal that would maintain its strength at that heat. In the end, they chose titanium with a covering of an incredibly tough, heat-resistant nickel-chromium alloy called Inconel X.

Like the Bell X-1, the X-15 was carried up by a mother-ship, a giant B-52 bomber, to an altitude of 40,000 feet. After being released, its powerful rocket engine fired for about 85 seconds, burning up to 15,000 pounds of fuel in that time, and pushing the plane and its pilot to accelerations of as much as 4g. From drop to landing, an entire flight would last about twelve minutes.

Among those who flew the X-15 was Neil Armstrong, the first man on the Moon. Less well known is Joseph Walker, a

**23** Joe Walker exiting his X-1A, cowboy style.

NASA test pilot who can claim another remarkable record – the first man to fly into space on two different occasions.

A veteran of the Second World War, Walker had been raised on a farm in Pennsylvania and showed an engineering talent and a thirst for knowledge early on. After the war, he joined NACA's Aircraft Engine Research Lab in Cleveland, Ohio as an experimental physicist, and later became a test pilot at the Edwards Flight Research Facility (now Dryden Flight Research Center) alongside Edwards Air Force Base. There he flew a variety of experimental aircraft, including the X-1 and its variants, and later X-planes.

In 1958, Walker was picked to take part in the US Air Force's Man In Space Soonest programme, aimed at putting an American in space before the Soviet Union. The programme was cancelled after a few months and replaced by NASA's Project Mercury. Only two men from the Man In Space Soonest programme would go on to reach space – Neil Armstrong and Walker. These two were also among the select few to pilot the Lunar Landing Research Vehicle (LLRV), a notoriously hard-to-fly contraption used to simulate the descent of the Apollo Lunar Module onto the surface of the Moon. On 30 October 1964, Walker was the first to lift off in the LLRV, taking it for three flights lasting a total of just under a minute to a peak altitude of ten feet.

In 1960, Walker became the first NASA pilot to fly the X-15, and the second X-15 pilot of all, after Scott Crossfield who ran initial tests on behalf of the manufacturer, North American Aviation. On Walker's maiden outing in the X-15, he was shocked by the plane's brutal acceleration. Having just ignited the engine he was crushed back into his seat, yelling 'Oh, my God!', to which the flight controller calmly replied, 'Yes? You called?'

Walker went on to fly the X-15 a couple of dozen times – twice in succession in 1963 to heights of more than 62 miles (100 kilometres) which, in the eyes of both the US Air Force and Fédération Aéronautique Internationale (FAI), counts as the edge of space. On 19 July he peaked out at 65.8 miles (105.9 kilometres), and a month later bettered that by going to 67 miles (107.8 kilometres), travelling at a speed of over 3,700 miles per hour on both occasions. The first of these flights

**24** Joe Walker after a flight of the X-15 #2.

made Walker the first US civilian in space and the second civilian astronaut in history after the Soviet Union's Valentina Tereshkova (who wasn't in the military at the time of her trip into orbit); the second flight made him the first person to go into space twice (according to Air Force and FAI rules). No manned plane would beat Walker's X-15 record altitude until Virgin Galactic's SpaceShipOne soared to over sixty-nine miles in 2004.

On 8 June 1966, Walker was flying an F-104 Starfighter in tight formation with other high-performance planes for a US Air Force publicity photo. He was trying to maintain position just below, and seventy feet to the side of, an XB-70 Valkyrie prototype bomber. At some point he misjudged where he was, and may also have flown into the wake vortex of the other plane. In any event, his Starfighter drifted too close to the XB-70's right wingtip, made contact with it, and flipped over. Rolling inverted, the Starfighter passed over the top of the bomber, smashed into its vertical stabilizers, and exploded, killing Walker instantly.

# 9

# HOSTILE SKIES
# AND AMAZING LEAPS

The prospect of making any kind of parachute jump would be daunting to most people. But doing it from twenty miles up is almost unthinkable. At that height, the curvature of the Earth is clearly visible, the sky is black even during the day, and a person would have to plunge for mile after mile, hurtling almost at the speed of sound, before the air got thick enough to slow their fall. As part of US Air Force investigations, Joseph William Kittinger II, a tough, wiry pilot (nicknamed 'Little Joe' because of his short stature), with a shock of orange hair and an engaging smile, made the highest, longest, fastest skydive in history. His record, finally broken in 2012, stood for more than half a century.

Flight was in Kittinger's blood from childhood. Born in 1928 in Tampa, Florida, he had his first plane ride when he was only eighteen months old, with his father aboard a Ford Trimotor, an all-metal three-engined plane. He grew up passionate about building model aircraft, never doubting he was going to be a pilot someday, and, when old enough, raced speedboats and learned how to fly in a Piper J-3 Cub. After two years' study at the University of Florida, he enlisted in the

US Air Force in 1949 and earned his pilot's wings the following year. His stock-in-trade approach was 'always volunteer' – even if you don't know what lies ahead. He volunteered to go to Korea to fight in the war that had just started there, but was assigned instead to Ramstein Air Base in West Germany.

On arrival in Europe, Kittinger started out flying P-47 Thunderbolts – one of the stalwart American fighter planes in the Second World War – before these were replaced at Ramstein by F-84 Thunderjets. In 1952 he volunteered for a stint working as a NATO test pilot in Copenhagen, flying a new, modified version of the Thunderjet called the F-84G, which could carry a nuclear bomb. The experience sparked Kittinger's interest in test flying, so that when the time came around for his reassignment in 1954, he applied to join the Research and Development Command at Holloman Air Force Base, near Alamogordo, New Mexico. It was a move that within a few years would lead Kittinger to the edge of space on one of the most extraordinary adventures in the annals of aviation.

## FROM HERO TO ZERO (G)

At Holloman, Kittinger flew an enthusiast's dream collection of experimental and high-performance planes, including the Mach 2 Lockheed F-104 Starfighter. Through his friendship with rocket-sled hero and aerospace medic John Stapp, he also got involved with testing how the environment of space would affect human beings.

Everyone knew in the mid-1950s that it wouldn't be long before people were blasted into Earth's orbit and beyond. But how would the human body react to weightlessness? Could astronauts survive for long, even wearing a special suit, if they had to escape from their capsule into a bone-chilling vacuum? Scientists needed ways to check these things before the first manned rockets went up.

Unfortunately, it is not possible to switch off gravity. But by flying a plane in a certain way, the effect of weightlessness can be produced for short periods. The technique was first proposed in the late 1940s by the German physicist Heinz Haber and his brother Fritz. As early as 1950, Chuck Yeager and Scott Crossfield, the famous supersonic test pilots of the X-planes, had used the Haber method to simulate zero-g. It involved power-diving an aircraft steeply, then pulling back into a hard climb before easing into a huge, precise parabolic arc at the top of the manoeuvre when, for a few precious moments, the strange and euphoric sensation of weightlessness would be felt. In the early days of supersonic flight, just about every top test pilot pulled the stunt at least once, for the heck of it and because it was fun, but no one had ever used it as a tool for serious research.

In 1953, Kittinger agreed to fly some zero-g parabolas at Holloman with physician David Simons alongside him to study the effects scientifically. Simons already had experience with human rocket-sled runs and some of the early launches of animals high into the atmosphere. Kittinger and Simons set off on their first zero-g mission in an F-89 Scorpion interceptor, and in no time Kittinger was flying perfect parabolas and stretching out the intervals of weightlessness to their maximum possible – around half a minute.

So mesmerized were the pair by the experience that they hardly noticed that two hours had gone by and their fuel was getting low. In fact, they were already eating into their reserve supply when Kittinger realized that not only did they need to head back to base straight away, but that they would have to request an emergency landing. Moments later, the situation became dire when the fuel booster pump on the number two engine failed and they were down to a single engine. Kittinger lined up the F-89 for landing but the tower refused permission because another jet was already approaching on an intersecting runway. 'Thanks for the information,' said Kittinger. 'I'm coming in.' It was then that he noticed something interesting about his landing gear. According to the indicator, he didn't have any – at least, nothing that had come down and locked. Banking sharply, he narrowly missed an incoming Sabrejet, and lined up on another runway. With his fuel gauge reading zero, he had no choice now but to touch down, landing gear or not. Mercifully, the indicator had lied – the wheels were in place – and Kittinger was able to bring the plane down safely. Seconds after landing, the remaining engine sputtered and died, the last drops of fuel having been spent. Grinning, Kittinger turned to Simons and said something that sounded like 'luck'.

## ELEVATOR TO SPACE

Planes will only fly so high, and it was clear that to learn more about how humans would react to the environment of space, subjects would have to travel much higher and for long periods. They would have to venture deep into the stratosphere for hours at a stretch so that careful physiological and psychological measurements could be made on them. The only way that was possible, in the days before the first manned rockets, was to use high-altitude balloons. In 1955 both Kittinger and

Simons were invited to take part in a new programme, called Project Manhigh, that would loft them above ninety-nine per cent of the atmosphere and into one of the most unfriendly realms known to man. Both volunteered without hesitation.

Before they were allowed to ascend on a Manhigh mission, however, they had plenty of preparation to do. Both men had to learn how to do parachute jumps, endure a simulated atmosphere of 100,000 feet in a decompression chamber, and, most terrifying of all, spend a full day shut up inside the incredibly cramped cylindrical metal gondola to be used in the project. Barely wider inside than a man's shoulders, and with only enough room for a pilot to stay seated in the same position, hour after hour, this pressurized, sealed vessel was the ultimate nightmare for anyone who suffered even a twinge of claustrophobia. To make the ordeal worse, the prospective pilot had to wear a partial-pressure suit of the type used by high-altitude jet fliers – an uncomfortable garment, made intentionally too small so that it would push blood up to the brain. The idea of the suit was that if the gondola sprang a leak, many miles above the ground, the oxygen inside the suit would squeeze the body as tightly as the cuff of a blood-pressure gauge and help keep the pilot conscious long enough to have a chance of descending to a safe altitude. Initially, Kittinger needed some counselling to overcome his anxiety of being so confined. But in the end both Simons and Kittinger successfully came through the claustrophobia and decompression tests, and passed the parachute training given to paratroopers. Now they were ready to take an elevator ride to the stars.

Kittinger was the first to go on the Manhigh I mission. Half an hour after midnight on 2 June 1957, he entered his capsule at the Winzen Research plant in Minneapolis where the capsule and balloon had been made, and was then sealed inside and taken by truck to Fleming Field in South St Paul for the launch. A giant, gossamer-thin balloon was inflated with helium, fluttering and gradually lifting off the ground as the light gas flowed in. At ground level, when ready to be released, the balloon was 200 feet high, and 35 to 40 feet wide, but would swell to 250 feet wide as the surrounding air pressure dropped during ascent. Six hours after entering his horribly confined quarters, Kittinger and his capsule finally headed up, bound for the stratosphere.

Twenty minutes into the flight, a problem with his radio gear meant that he could no longer send voice messages, only receive them, so he had to use Morse code for all his responses. An hour and a half after taking off, the balloon reached its maximum height above sea level of 95,200 feet – a new human altitude record. But then more trouble came along. A fault in the cabin pressure regulator led to Kittinger's oxygen supply running down much faster than planned. A message from ground control said that the flight needed to be cut short and it was time to come home. Kittinger replied by tapping out 'C-O-M-E-A-N-D-G-E-T-M-E'. Simons, who was monitoring Kittinger for signs of physical or mental distress, thought the pilot was experiencing what was known as 'breakaway phenomenon' – a kind of hallucination or disorientation, reported by some high-altitude pilots, of feeling detached from the Earth. But there was nothing wrong with Kittinger. He was just joking, although he did admit later that if he had known there were concerns about 'breakaway' he would never have made the remark. The descent went smoothly and, just over twelve hours after launch, Manhigh

I ended its journey safely when the capsule set down in a clearing on the bank of a creek, sixty-five miles or so northeast of its starting point.

Five months later, Simons flew in Manhigh II and set a new altitude record, reaching 101,500 feet on a flight that lasted a mind-and-body-testing thirty-two hours. The following year, Manhigh III, the last of the series, took another volunteer, Air Force Lieutenant Clifton McClure, up to almost 100,000 feet.

## FALLS THROUGH THE STRATOSPHERE

For Kittinger, Manhigh was just the prelude to an even more extraordinary set of balloon trips to the frontier of space. A new generation of super-high-flying aircraft was coming into service, including the U-2 spy plane, which could cruise at extreme altitudes. The US Air Force was concerned about the survival of pilots who might have to bail out if anything went wrong so far above the ground. Tests with manikins had shown that, where the air is very thin, a person falling could go into a flat spin of more than 400 revolutions per minute, resulting in almost immediate loss of consciousness. New types of parachutes needed to be tested, which stabilized the wearer while in a near-vacuum free fall from many miles up, and deployed in stages as the air gradually grew denser. Stapp was put in charge of a new balloon research programme, called Project Excelsior, at the Aerospace Medical Research Labs at Wright-Patterson Air Force Base in Dayton, Ohio.

Unlike Manhigh, in which the pilots had ascended in a closed, sealed capsule with no way out until they got back to the ground, Excelsior would involve open gondolas so that, at the appointed time and height, the seated occupant, with the parachute system on his back, would simply lean forward and plunge towards the Earth, many miles below. If Manhigh was a poor choice for claustrophobics, then Excelsior was the thing to avoid if you had a problem with heights. Inches away from the tip of your boots was a sheer drop into nothingness – black airlessness – and temperatures of 110 degrees below zero. To survive those conditions, Kittinger had to wear a full pressure suit plus extra layers of thermal clothing, which together with the parachute system almost doubled his weight.

Kittinger made two successful jumps during the Excelsior I and II missions, from heights of about 75,000 feet, although the first almost ended in disaster. On reaching the altitude at which he was supposed to bail out, Kittinger found he couldn't get out of his seat because the bottles containing his water supply had frozen and expanded and wedged him in place. After eleven seconds he managed to break free, but in the struggle inadvertently set off the timer for his stabilization parachute. This caused the pilot chute to deploy before he had gained enough speed from his fall to properly activate the rest of the system. The end result was that the main parachute wrapped around Kittinger's neck and he couldn't get it disentangled. By now he was spinning out of control, twice around every second, pulling up to 22g at his extremities. With the blood draining from his head, he lost consciousness and would certainly have been killed had it not been for a handy little gadget contained in his equipment. An automatic parachute opener, triggered when the atmospheric pressure reached a certain level, opened the reserve chute at an elevation of 11,000 feet. One thousand feet later the canopy was fully open, allowing Kittinger to float down safely to the New Mexico desert.

The first two Excelsior jumps completed, it was now time for the big one. On 16 August 1960, Kittinger got up with the rest of the balloon crew at about 2 a.m. to start filling the helium balloon that would send him aloft. At 4 a.m. he started breathing in pure oxygen for two hours in order to flush out all the dissolved nitrogen from his blood so that he didn't get the dreaded bends – decompression sickness – due to the rapid ascent he was about to make. Then he started to dress for the ordeal, first putting on layers of warm clothing to insulate him from the bone-chilling exposure to extreme sub-zero temperatures in his exposed gondola, and finally a thin pressure suit and a sealed helmet supplied with supplemental oxygen. Until launch, he was kept in air-conditioned surroundings to avoid sweating, because any body moisture would have frozen solid on the way up.

With the helium balloon fully inflated to its sea-level pressure, it was released and Kittinger was under way. As he rose, just as on the earlier two missions, what struck him first was the numbing cold. Then part of his pressure suit failed: at 40,000 feet he realized that the glove on his right hand hadn't inflated. As the air pressure around his unprotected hand fell to almost nothing, his hand quickly swelled to twice its normal size and became useless for the rest of the flight. If he told the doctor on the ground about it, he knew he would be recalled and there was no guarantee that there would be another flight. So he kept quiet, and fortunately the rest of the pressure suit worked fine.

After an hour and a half, the balloon reached its ceiling – its maximum altitude – and it was time for him to step out. Kittinger went through his 46-step checklist then sent one last radio message to those on the ground: 'There is a hostile sky above me. Man will never conquer space. He may live in it, but he will never conquer it.' Then he disconnected himself from the balloon's power supply, severing all communication with his colleagues. He was now totally on his own, 102,800 feet (almost twenty miles) up – further than any human being had ever been from the Earth. All that stood between him and an instant death was his pressure suit and the kit on his back.

From the doorway of his gondola he could see below for 400 miles in every direction and far, far beneath him a deck of brilliant white clouds. Overhead it was inky black, but with no sign of stars because the glare from the Sun narrows the pupils too much to be able to see them. All he could hear was the sound of his own breathing.

With one final glance at the view, Kittinger jumped over the side and rolled over so that he was looking back up at the balloon, which seemed to be hurtling into space. In fact it was Kittinger who was hurtling – downwards – his speed building at a fantastic rate. Passing 90,000 feet he reached 614 miles per hour (not quite breaking the speed of sound, but coming close). The altimeter on his wrist was unwinding at a dizzy rate, yet he had no sensation of travelling so fast because there was no rush of air (it being so thin) and no passing landmarks by which to gauge his motion.

After plunging for just over four a half minutes, his main chute opened and he was back on solid ground eight minutes later, surrounded by his ecstatic crew. He smashed a fistful of world records that day: the highest ascent, the longest free fall, the longest parachute descent, and the fastest speed by a person, outside of a vehicle, through the atmosphere.

**25** Joseph Kittinger's record-breaking skydive from 102,800 feet (31,300 metres).

## OPEN SKIES

After Excelsior, the rest of Kittinger's career seems almost tame by comparison, although any of a string of his future exploits would have been the high point, literally, in the lives of most people. Back at Holloman Air Force Base, Kittinger took part in another high-altitude experiment called Project Stargazer. In December 1963, he and astronomer William C. White rode an open-gondola helium balloon packed with scientific equipment to a height of 82,200 feet, way above any atmospheric turbulence, and spent more than eighteen hours carrying out observations of the heavens. An early predecessor of the Hubble Space Telescope, Stargazer pushed the technological envelope of its day, especially in the stabilization system it used to keep the telescope steady, and involved the heaviest payload balloon ever built.

After a decade doing scientific research, Kittinger returned to his day job as a fighter pilot and served three combat tours of duty during the Vietnam War. Just a week before the end of his third tour, in May 1972, his F-4D Phantom was hit by an MiG-27 over North Vietnam, and Kittinger and his weapons officer, William Reich, were forced to eject. Landing near a village, they were set upon by an angry crowd. 'Within the first minute,' he recalled later, 'I was almost killed twice – once by an old lady and once by a kid with a machete.' The pair spent eleven months, enduring torture, in the infamous 'Hanoi Hilton' POW camp, in a cell next to senator-to-be John McCain. Kittinger kept up his morale by planning a round-the-world balloon flight.

After his release Kittinger remained in the US Air Force until he retired, as a colonel, in 1978. Ever since, he has been

**26** Joe Kittinger and the recovery crew following his record-breaking jump.

involved with ballooning, barnstorming, and other kinds of aerial activity, breaking several distance records for different classes of gas balloon. Most recently, he was on the team that helped Austrian Felix Baumgartner achieve a new highest skydive, beating his own record by four miles.

# KITTINGER AND
# BAUMGARTNER'S JUMPS COMPARED

| | KITTINGER | BAUMGARTNER |
|---|---|---|
| **AGE AT TIME OF JUMP** | 32 | 43 |
| **OCCUPATION** | Captain, US Air Force | Former Austrian military parachutist |
| **PREVIOUS PARACHUTE JUMPS** | 32 | 2,500 |
| **HEIGHT OF JUMP** | 102,800 feet (19.5 miles) | 128,100 feet (24.3 miles) |
| **MAXIMUM SPEED OF FALL** | 614 mph (Mach 0.9) | 834 mph (Mach 1.24) |
| **ASCENT VEHICLE** | Unpressurized gondola | Pressurized capsule |
| **BALLOON** | 3 million-cubic-foot helium balloon, 184 feet tall | 30-million-cubic-foot helium balloon, 335 feet tall |
| **ATTIRE** | Air Force standard partial pressure suit | Custom-made pressure suit |
| **BACKING ORGANIZATION** | US Air Force | Red Bull, energy-drink maker |

# 10

## TESTED TO
## THE EXTREME

'Sadistic' and 'diabolical' – just a couple of the terms used by past astronauts to describe the most intimidating piece of apparatus they encountered in their training. *Time* magazine described it as 'a gruesome merry-go-round'. The object in question was the Johnsville human centrifuge, the largest and most powerful machine of its kind on the planet.

Some of the people who rode most often in the Johnsville centrifuge, and subjected themselves to the most terrifying ordeals, weren't celebrities like astronauts John Glenn or Alan Shepard. They were little-known physicians, engineers, and technicians – back-room guys – whose job it was to pave the way for tests on the pilots who would actually fly into space. Among these unsung volunteers were the medical technician Art Guntner and the psychologist R. Flanagan Gray, inventor of the 'Iron Maiden', an accessory to the infamous human centrifuge named, appropriately, after a medieval torture device.

Before the start of the Space Age, no one knew if the human body could stand up to the harsh treatment it would receive getting into and out of space, not to mention what the environment of space itself would do to a man or woman. In

**27** The Johnsville centrifuge.

being blasted into orbit on top of a powerful rocket, an astronaut would feel, for minutes at a stretch, much heavier than on Earth. During re-entry, the g-forces would be even higher. Through experiments on the ground and the experiences of pilots during extreme aircraft manoeuvres, it was known that people could withstand brief exposure to very high g-force. But how much could space travellers take over longer periods, while accelerating into orbit and decelerating upon their return to Earth? And would they still be able to operate their vehicles?

## WHIRLED FIRST

Early in the US space programme, the idea was entertained of using amusement-park rides, including roller-coasters and various spinning attractions, to study the effects on people of sustained acceleration. But it soon became obvious that only purpose-built machines would be able to deliver the intense and unusual types of stress that researchers wanted to study. Nothing less than a human-rated centrifuge would do – and that was just for starters. A whole demented playground of devices would eventually be needed to test future spacefarers to the limits of their endurance.

The first reasonably powerful human centrifuge had been built during the Second World War to help develop an effective anti-g suit. Canadian scientists Wilbur Franks and Frederick Banting, whose wartime research involved the problem of pilots blacking out or suffering G-LOC (g-induced loss of consciousness) when pulling high g-force, realized that testing new suits in real aircraft wasn't an option. It was too dangerous, the experiment couldn't be precisely controlled, and there was a security risk – as the g-suit project was top-secret. The human centrifuge, set up at the Canadian military's Clinical Investigation Unit (CIU) in Toronto, was the perfect solution. It went into operation in mid-1941 and was supposed to be very hush-hush. However, it was pretty clear to everyone that something unusual was going on in the building where the device was housed. The 200-horsepower motor that drove it shared a power line with the rest of the neighbourhood, and every time it was fired up it drained current from the overhead lines supplying streetcars on the road outside so that they ground to a halt.

The CIU centrifuge had a round gondola hung from a horizontal arm that was attached to a vertical shaft. The motor spun the shaft, causing the gondola to swing out so that it was tilted at almost ninety degrees when moving fast – an arrangement familiar to anyone who has been on the fairground ride known as 'the Enterprise'. The seat inside was like that of a fighter plane and suspended independently of the gondola, so that the rider could be positioned at different angles – even upside down to produce negative g.

An observer in the control room sent signals into the gondola by turning on lights and sounding a buzzer; the rider replied by switching the signals off. If he turned off only the buzzer it meant he was still conscious but had suffered a blackout – a total loss of vision. If he failed to turn off the lights or the buzzer, it was the sign that he had gone unconscious.

The results of the CIU centrifuge helped Franks to develop the first operationally practical g-suit. But the machine was also used to evaluate aircrew trainees – especially those suspected by their instructors of having a low tolerance to g loads. Quite possibly it saved a few lives by identifying unsuitable candidates who could then be moved out of pilot training before they got into trouble in the air.

## RIDING THE WHEEL

Something on an altogether larger scale, though, was needed to prepare for America's manned exploits in space. The Johnsville human centrifuge, at the Naval Air Development Center (NADC) in Warminster, Pennsylvania, just outside Philadelphia, would play a key role in testing astronauts for the Mercury, Gemini, and Apollo programmes. But its first objective, when it was completed in July 1950, was to study the effects of g-forces produced by high-performance aircraft and, in particular, to help train pilots for flying the X-15.

'The Wheel', as it was not-so-affectionately known, was housed within a round, cavernous, 11,000-square-foot building at the NADC (later renamed the Naval Air Warfare Center), a facility where the US Navy also modified aircraft before they joined the fleet and carried out weapons research. The site was chosen because the turning forces that the centrifuge could produce were so strong that the giant machine had to be anchored directly to bedrock to stop it shaking itself loose, and Warminster had some of the most stable bedrock that engineers could find.

The steel gondola of the centrifuge was shaped like a flattened sphere, measuring ten feet by six feet, and was mounted on a fifty-foot arm, at the other end of which was a 4,000-horsepower General Electric engine. So powerful was this motor that, flat out, it could whip the gondola to speeds of 175 miles per hour in just seven seconds, reaching a potentially deadly maximum of 40g. Equipped with dual gimbals, the gondola could be rotated so that the test subject was oriented in various positions relative to the applied g-force. Simply by turning the gondola as it spun about the arm, experimenters could subject pilots to positive g-force (with acceleration in a head-to-foot direction), negative g-force (foot-to-head, like that felt in a rapidly descending lift), transverse (chest to back), lateral (side to side), and practically every other variation that could crop up during a mission.

Naturally, it is the great test pilots and astronauts whose experiences on the centrifuge tend to be most heard about. But a lot of the time, the men who rode 'the Wheel' were ordinary staff members at Johnsville or other US Navy personnel who put themselves forward. The big names of aerospace were too busy flying to spend hour after hour being research guinea pigs, especially in the early days of the Wheel when its performance was being evaluated. That job fell instead to local volunteers such as aerospace medical technician Art Guntner who, in his time at NADC, climbed into the belly of the beast some 350 times and briefly soaked up to 15g of punishment – an acceleration that dwarfs the 4g to 6g max experienced aboard top-fuel dragsters, high-g roller-coasters, or even the Space Shuttle.

Born and raised in Morgantown, West Virginia, and hailing from a family with a long military tradition, Guntner joined the US Navy in 1958 and graduated from Aerospace Medicine School in 1960. Straight after that he was sent to NADC to work at the Aviation Medical Acceleration Laboratory, where the giant centrifuge was located. This was around the time that the first American astronauts-to-be were scheduled to begin tests on the machine.

The men – and they were all men (women having been excluded from the selection process) – chosen to take part in America's first human space flights were known as the Mercury Seven. They were the elite corps who would fly aboard the cramped Mercury capsules that would be launched into space by Redstone and Atlas rockets.

It was Guntner's job to prep the Mercury Seven before they rode in the centrifuge. He helped brief them on the results of the early simulations and his own experiences at high-g. Another part of his job was to attach sensors to the astronauts before they were sealed inside the gondola, to monitor their heart rate and breathing. The gondola was kitted out with an instrument panel and hand controller just like that in the Mercury capsule, and could be depressurized to the actual flight pressure of five pounds per square inch. Test subjects

# THE MERCURY SEVEN

The Mercury Seven were the American military test pilots selected on 9 April 1959 to take part in the Mercury Project – the first US manned space programme. They included Scott Carpenter, Gordon Cooper, John Glenn, Virgil 'Gus' Grissom, Walter Schirra, Alan Shepard, and Donald ('Deke') Slayton. All but Slayton actually flew on a Mercury mission.

The top age for candidate Mercury astronauts was set at forty, the maximum height at 5 feet 11 inches (1.80 metres), and the maximum weight at 180 pounds (81.6 kilogrammes). The selection of the final seven, from 508 pilots originally deemed to meet the basic astronaut requirements, came on 9 April 1959. Shepard was the tallest Mercury astronaut at 5 feet 11 inches (1.80 metres), Grissom the shortest at 5 feet 7 inches (1.70 metres). Cooper was the youngest at thirty-two, Glenn the oldest at thirty-seven. The first message received by the test pilots as to their impending call for astronaut service read: 'You will soon receive orders to OP-05 in Washington in connection with a special project. Please do not discuss the matter with anyone or speculate on the purpose of the orders, as any prior identification of yourself with the project might prejudice that project.'

**28** The Mercury Seven astronauts with a model of an Atlas rocket. Standing, left to right, are Alan B. Shepard Jr, Walter M. Schirra Jr, and John H. Glenn Jr; sitting, left to right, are Virgil I. Grissom, M. Scott Carpenter, Donald ('Deke') Slayton, and L. Gordon Cooper Jr.

would then report on how well they could use the controls while under high-g loading, and describe any adverse effects they felt. An especially valuable but frightening feature of the gondola was that it could be put into a tumble in which the accelerations might lurch gut-wrenchingly from positive 10g to negative 10g in roughly one second.

But there were lighter moments amid the stress and hard work. One weekend, Guntner was called in at short notice to work with the astronauts and, as his wife had other commitments, he had to bring his two-year-old daughter along with him. It proved to be a long, busy day with only a skeleton staff on duty, so Guntner had to rely on the only babysitters available – Alan Shepard and John Glenn.

For the Mercury astronauts, and, later, Gemini and Apollo astronauts, the Wheel was both a rite of passage and an essential training tool. 'Whirling around at the end of that long arm, I was acting as a guinea pig for what a human being might encounter being launched into space or re-entering the atmosphere,' John Glenn recalled in his autobiography. 'You were straining every muscle of your body to the maximum … if you even thought of easing up, your vision would narrow like a set of blinkers and you'd start to black out.'

Under the crushing force of 6g, 8g, or even 10g, normal breathing isn't an option. Exhaling is no problem but inhaling in the usual way, when it feels as if you weigh half a ton or more, is out of the question. '[I]t's impossible to reinflate your lungs,' wrote Apollo 11 Command Module pilot Michael Collins, 'just as if steel bands were tightly encircling your chest. So you have to develop an entirely new method, keeping the lungs almost fully inflated at all times, and giving rapid little pants "off the top".'

**29** Alan Shepard poised on the step of the Johnsville centrifuge prior to a training run.

Subjects rode the centrifuge in one of two modes: closed-loop, in which the individual had full control over the movements of the gondola; and open-loop, or 'meat in a seat', in which the rider was more or less a lab rat. In every case, doctors monitored the progress of the 'flight', moment by moment, and both the subject and the support team could stop the ride immediately if there was a concern.

## PUSHING THE ENVELOPE

Before anyone was allowed aboard the gondola for their first spin, they had to go through a battery of medical tests. The organ of most concern was the heart because it was entirely possible for someone to have an unsuspected heart condition, such as an arrhythmia, that might prove life-threatening if triggered by the stress of high g-force. In fact, an undiagnosed arrhythmia was triggered when Donald ('Deke') Slayton took a high-g ride, disqualifying him from astronaut duty in the Mercury Project (although he did eventually make it into space with the 1975 Apollo-Soyuz Test Project).

Those given the green light to ride varied a lot in how well they tolerated the g-forces of the Wheel. Some people not only coped well with the centrifuge, but strove to test its limits – and their own. One of these hardy souls was US Navy Reserve officer Carter Collins who, in August 1958, withstood a crushing 20g for six seconds, using a grunting technique to avoid blacking out and suffering chest pains. Later that day, R. Flanagan Gray, a civilian psychologist, repeated the feat.

Gray was one of the researchers at Johnsville trying to come up with new ways for astronauts to cope with high g-forces during take-off and re-entry. Some of the fruits of this work, such as contoured couches and special restraining straps, found their way into the actual spacecraft design; other ideas, although effective, weren't used for practical reasons.

A lot of Gray's experiments involved using animals instead of people and wouldn't be allowed today. At the time, though, they would have seemed acceptable and Gray was known by his family as a kind and compassionate man. He would often bring home the animals used in his high-g tests to see if they could adapt as pets and live out the rest of their lives free of the trauma they'd had to endure at Johnsville. But it didn't usually work out. On one occasion he brought back a goldfish after a multi-g whirl. It could only manage to swim upside down in its new home for a couple of days before it expired. It had been one of the earliest subjects to take a trip aboard one of Gray's most memorable inventions – something menacingly called the 'Iron Maiden'.

Early research with animals had shown that being immersed in water helped dissipate g-forces. Would it work with humans? To find out, Gray designed an aluminium capsule, sculpted roughly in the shape of a seated person, which could be filled with water and then attached to the centrifuge, giving the effect of being in a strange bathtub caught in a violent whirlpool. Into this, in March 1958, Gray climbed before being immersed up to his ribs and then spun around, lying down, head pointing in the direction of motion. On this first, semi-submerged trip, he took a full 16g without harm. The following year he entered the Iron Maiden again, for the ride of his life. This time he was seated facing backwards and totally immersed, at the last moment, so that the water was over the top of his head. Then, in an ordeal of which Harry Houdini would have

**30** The 'Iron Maiden', a device patented by R. Flannigan Gray.

been proud, he held his breath for twenty-five seconds while being spun crazily to a five-second peak of 32g – the most ever endured by a human in a centrifuge. Throughout the whole manic experience Gray remained conscious, although he did not emerge unscathed. He reported some sinus pain, but more seriously he received a hernia and suffered an eye injury that permanently affected his vision. After the ordeal, he

## LA-Z-BOY TO MARS

One of the strangest experiments using the Johnsville centrifuge stemmed from a problem that two scientists, Carl Clark and James Hardy, posed in the late 1950s. If a spacecraft could accelerate steadily at 2g (before slowing down again at the other end) it would be able to get to Mars, for instance, in just a few days. Setting aside the obvious propulsion issues, Clark and Hardy were interested in what effect that level of moderate but constant g-loading might have on a person. And what better way to find out than by putting yourself in the hot seat? In fact, Clark brought along his very own seat – a La-Z-Boy recliner – from home, set it up in the gondola, and just moved in for a whole day. With the centrifuge whirring around, producing a steady two Earth gravities, Clark worked, ate, and slept, over a twenty-four-hour period inside his little high-g world. At the end of it the only adverse effect he had was mild fatigue.

was driven home in a black government car and could barely walk into his house.

Astonishingly, he was ready to take more but the experimental set-up couldn't handle it. The Wheel could deliver up to 40g, but only at the end of its arm, inside the gondola. The Iron Maiden was too big to fit inside the gondola and so had to be mounted closer to the central axis, where 32g was the highest acceleration possible.

## THE SCREAM MACHINE

Only one other simulator rivalled the Wheel for extreme unpleasantness in the eyes of the Mercury Seven. Called the MASTIF (Multiple Axis Space Test Inertia Facility), it was located at NASA's Lewis (now Glenn) Research Center in Cleveland, Ohio, and looked like something that a talented giant had made from a Tinker Toy set. It consisted of three concentric cages gimballed so that each could rotate about a different axis at right angles to the other two. At its core was a yoke that was twenty-one feet across that could support a 3,000-pound space capsule, which could be tumbled and turned in a noisy, incredibly dizzying way at up to sixty revolutions per minute. Its first job was to test the attitude control system of the Mercury capsule used for a flight called Big Joe 1 – an unmanned mission involving an Atlas booster to check out the capsule's heat shield during re-entry.

An engineer at Lewis, James Useller, saw the potential for the MASTIF to be used for training astronauts in emergency situations. The unmanned capsule was replaced with a seat that was equipped on the arm rests with two hand controls linked to nitrogen thrusters. With the rig tumbling this way and that, the astronaut had the task of trying to right the device and keep it stable.

In mid-1959, Useller and a Lewis test pilot, Joseph Algranti, began taking cautious rides inside the MASTIF. Then they set up a formal test programme for about ten pilots and research scientists who wanted to see what effect rolling, pitching, and yawing at different speeds and for different lengths of time would have on a person. Beyond about thirty revs per minute, whirling around in three axes at once, everyone – including the most experienced pilots – suffered motion sickness. All was now ready for the Mercury Seven to take the fun ride.

In February 1960, the first pair of Mercury astronauts, Gus Grissom and Alan Shepard, arrived in Cleveland for a week's stay to test the MASTIF and their reactions to it. After a long

**31** The MASTIF (Multiple Axis Space Test Inertia Facility) at Lewis Research Center in 1960.

previous session in which Shepard was strapped stationary in the seat of terror while the rig was made ready, he climbed back in for his first nausea-inducing spin. Within seconds of the MASTIF going into its horrible gyrations, an ashen-faced Shepard hit the red 'chicken switch', sounding a claxon horn as a signal to stop the ordeal. The next day Shepard – and before the end of March all the astronauts – endured runs at the full thirty revolutions per minute in all three axes and quickly learned, by using the hand controllers, to activate the thrusters, to halt their rotation and bring themselves to a stop while the cages continued to spin. The skill they picked up may have proved invaluable in an emergency, but one series on the MASTIF was enough. Reporters who watched a demonstration by Scott Carpenter afterwards gave vivid descriptions of the hapless astronaut's piercing scream and the extraordinary contortions of this 'ultimate' in wild carnival rides.

# 11

## BREAK-UP
## AT MACH 3

No day is routine if you're flying a plane such as the Lockheed SR-71 Blackbird. Designed to spy on enemy territory from heights of up to 85,000 feet and to reach speeds greater than Mach 3, its standard plan if attacked by a surface-to-air missile was not to try to dodge or destroy the threat but simply to outrun it. Every mission in a Blackbird was an adventure, but none was as utterly incredible as that experienced by Bill Weaver, who suddenly woke to find himself sailing through the atmosphere at high speed – having somehow ended up outside his aircraft.

It was 25 January 1966. William A. Weaver, Lockheed's chief test pilot, and James T. Zwayer, one of the company's flight-test specialists, were scheduled to fly their Blackbird on a mission out of Edwards Air Force Base in California's Mojave Desert. Earlier that month the first SR-71 to enter service with the US Air Force had been delivered to Beale Air Force Base in northern California. It was still a very new plane – flown for the first time only just over a year earlier – and there was much to be learned about how to improve its performance.

## ANATOMY OF A BLACKBIRD

A product of Cold War strategy, the Blackbird was intended to complement America's other spy plane – the slender, narrow-winged U-2, one of which had been famously shot down over the Soviet Union in 1960 with the capture of its pilot, Gary Powers. The Blackbird was the first plane to use stealth technology, in its shape and materials, making it hard to spot on radar; it was also more than four times faster than the U-2, making it almost impossible to intercept. Like the U-2, it was a black project developed under the guidance of legendary aircraft engineer and designer Clarence 'Kelly' Johnson at Lockheed's secretive plant known as the 'Skunk Works'.

One of the problems with flying at triple the speed of sound – about 2,200 miles per hour – is that the plane's wings and fuselage get very hot. To combat this, the designers of the SR-71 used materials that could easily withstand high temperatures. Most of the airframe was built from titanium, ironically imported from the USSR at the height of the Cold War, the Soviets having been led astray about the purpose for the metal. After construction, the plane was painted a very dark blue – almost black (hence its nickname) – to help radiate away heat and also act as a camouflage, since the aircraft did its reconnaissance at night.

Another major innovation on the SR-71 was its giant Pratt & Whitney J58-P4 engines. One on each side, with its own upright tail, they seemed to dwarf the stubby triangular wings into which they were built. No engines before had been designed to run continuously on afterburner, but then again no engines had ever previously been made that ran most efficiently at Mach 3.2 and could operate as either a turbojet or

# LOCKHEED SR-71 BLACKBIRD

**Crew:** 2 (pilot and reconnaissance systems officer)
**Length:** 107 ft 5 in (32.74 m)
**Wingspan:** 55 ft 7 in (16.95 m)
**Empty weight:** 67,500 lbs (30,600 kgs)
**Powerplant:** 2x Pratt & Whitney J58-1 turbojets
**Maximum speed:** Mach 3.3 (2,200 + mph)
**Service Ceiling:** 85,000 ft (25,900 m)

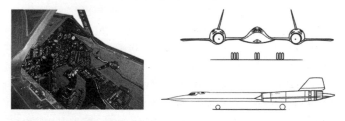

**32** An SR-71 Blackbird flies over the snow-covered southern Sierra Nevada Mountains of California after being refuelled by a US Air Force tanker during a 1994 flight.

a ramjet. At lower speeds, in turbojet mode, the J58s gulped down fuel like there was no tomorrow – and the special JP-7 jet fuel used by the Blackbird cost roughly as much as vintage Scotch. A full 12,000-gallon (38-ton) tank-load of fuel was eye-wateringly expensive and consumed so quickly that the SR-71 had to be resupplied every couple of hours on the wing by a tanker plane.

Crucial to the proper working of the Blackbird's engines was a sharp cone, known as the spike, which was smack in the middle of each air intake. On the ground and at subsonic speeds this was locked in its full forward position, but as the plane slipped past Mach 1.6 the spike withdrew further and further into the cowling. As the spike moved back, the inlet bypass doors, in the side of the nacelle (the streamlined cover around the engine), closed to regulate the flow of air around the engine. All this was handled automatically, from millisecond to millisecond, the purpose being to make sure that, however fast the aircraft was travelling, the air coming into the engine intake was slowed to subsonic speed before it reached the front of the compressor.

When air goes from supersonic to subsonic speeds it forms a shock wave. Inside the Blackbird's inlets this shock wave had to be kept in just the right position to ensure the engine ran smoothly and at peak performance. If something went wrong with the control of the air flow, the shock wave could shoot forwards, blocking any air from entering the inlet, in an event known as inlet unstart. In a split second the affected engine would lose all thrust, there would be explosive banging noises, and the plane would start to swing violently from side to side. SR-71 pilots who had experienced unstarts said that the yawing was so bad it was like the nose and tail were trying to swap places. It could be severe enough to crack a helmet as the pilot's head was smashed into the cockpit canopy by the lurching. Usually, the plane's auto systems kicked in straight away, making sure the shock wave was brought back in check and that normal service resumed.

## ANOTHER DAY AT 80,000 FEET

Preparations for Weaver and Zwayer's flight began early in the morning; preflight checks of the Blackbird were never quick affairs. Finally, at 11.20 a.m., all was ready and DUTCH 54, as the mission was called, rolled out onto the runway. Among the goals of the mission were to carry out tests on some aspects of the plane's reconnaissance sensors and navigation system. The crew were also scheduled to check out some procedures aimed at cutting drag and improving performance at high Mach speeds. These procedures involved flying with the Blackbird's centre of gravity shifted further to the rear of the plane than normal – a configuration that, Weaver noticed early on in the flight, tended to throw off the Auto-Nav (the onboard computer navigation system).

Take-off, climb, and acceleration were all normal – that is, for an aircraft that can get 12,000 feet off the ground in about a minute. The first round of tests went well and the plane dropped in altitude so that it could refuel from its KC-135 tanker. DUTCH 54 then headed eastward and up. Nothing on a Blackbird test flight was left to chance or pilot discretion. Every mission, every manoeuvre of the plane, was planned out meticulously in advance. Unless something unforeseen arose, the pilot followed the profile laid out for him to the

**33** NASA's SR-71 taking off from Dryden Flight Research Center.

letter. The next stage in the prearranged plan was to go into
a banked, thirty-five-degree turn to the right before peaking
at Mach 3.2 and levelling out at an initial cruise altitude of
around 80,000 feet.

Weaver went into the planned turn at around Mach 3.17
and 78,000 feet, somewhere over the little town of Albert,
New Mexico, boosting engine air flow to counter any drop
in altitude. That's when the trouble started. As soon as he
started to bank, there was an unstart in the starboard engine
signalled by a loud bang. With the other engine still thrusting, the plane immediately rolled further right and at the
same time the nose pitched up. To try to compensate, Weaver
yanked the control stick forward and to the left as far as it
would go, but it had no effect. A couple of seconds after the

unstart, the plane was catastrophically out of control – and about to break up.

## THE IMPOSSIBLE ESCAPE

Weaver knew that they should eject but reckoned their chances of surviving fifteen miles up while hurtling along at three times the speed of sound, without anything but a pressure suit, were pretty close to zero. He yelled over the intercom to Zwayer that they should stay in the plane until it dropped to a lower speed and altitude, but the turn was so steep and the g-force from it so intense that his words came out garbled and unintelligible.

Weaver's last recollection, a moment before losing consciousness, was of the Blackbird disintegrating around him. If anyone could have seen what happened next, they would have abandoned all hope for the survival of the crew. The fuselage snapped, just ahead of where the wings joined. The front part of the plane, containing the crew compartments, bent to one side, allowing Mach 3 air to rush into and around the exposed rear section, stripping away panels and ripping apart wings, engine nacelles and the rest, causing a rain of high-tech debris to fall towards the arid (and, fortunately, mostly deserted) ranch lands below.

On coming round, Weaver's first thoughts were as muddled as his vision. Was he dreaming? Was he dead? If this was death, he thought, it was pretty good. He felt detached – which he was, in more ways than one! – and even euphoric. But as reason returned, it became clear to him that, by some amazing fluke, he was still alive. Somehow, without actually ejecting, he had been thrown clear of the plane when it came apart and he had survived – so far.

He certainly wouldn't have survived if his pressure suit had deflated, which meant the emergency oxygen cylinder attached to his parachute harness was working. Above 62,000 feet, without pressurization, blood begins to boil inside the body – for the same reason that the boiling point of water is much lower on tall mountains – but not before all the air is forcefully driven out of the lungs, causing suffocation. The inflated suit had also served as a personal escape pod, helping protect Weaver from some of the vicious battering and jarring that he would otherwise have been subjected to when he was wrenched out of his cockpit, and, of course, it had given him something to breathe. Another piece of luck was that a small stabilizing parachute had deployed, even though Weaver hadn't gone through a proper ejection. This small chute had prevented him from spinning around wildly in the low-density air at high altitude.

So much for the good news. The bad news was that he was many thousands of feet up, and falling rapidly. He could hear the rush of air going past him and what sounded like straps flapping in the wind. But he couldn't see anything other than a white haze because the face plate of his helmet was completely iced over. He had no visual clues to tell him how high up he was and he didn't know how long he'd been unconscious. His main parachute was supposed to open automatically at 15,000 feet, but had it been damaged in the chaos of recent events?

He decided to try to open his main chute by hand and fumbled for the D-ring that would activate it. What would normally have been an easy action became virtually impossible in the inflated and inflexible pressure suit, working blind, and with fingers numbed by the extreme cold. Weaver couldn't find the ring and instead decided to lift his faceplate so that he could at least get some idea of his height above the ground. Just as he reached to do this he felt the sudden upward jerk of his main parachute deploying – the automatic system had done its job. Holding his faceplate open with one hand, because the retaining latch had broken, he saw clear blue sky and, below, miles and miles of desolate open country, covered here and there with patches of winter snow but no signs of habitation. Smoke and fire, rising in some places, told him where large pieces of wreckage from the Blackbird had come down. About a quarter of a mile away, he was heartened to see Zwayer's parachute descending. He, too, had been freed from the destruction when the plane's crew canopy was torn away – but, it turned out, with very different results.

It was now 3 p.m. and, being the middle of winter, it would be turning dark in a few hours – and bitterly cold. Weaver started thinking about survival on the ground. Search-and-rescue crews would soon be out looking but it might be some time before he and Zwayer were found.

But first there was the little matter of getting down safely. What a cruel twist of fate it would be if he escaped virtually unhurt from a disintegrating plane at Mach 3 only to break a leg or worse on landing – and even expert pilots can be novices when it comes to parachute jumps. This happened to be Weaver's first, and it was no easy one. He lacked any feeling in his hands, from their sub-zero exposure, and had to use one of them to hold up his faceplate, which meant he couldn't steer or rotate the chute by pulling on the risers.

About 300 feet above the ground, he pulled the release handle on his survival kit, making sure that it was still tied to him by a slack lanyard. Mentally he went over what was in the kit that he might soon be needing. Then, the landing. It turned out to be surprisingly smooth, on soft level ground, clear of any painful obstacles like rocks or cacti.

## AFTERMATH

In the few minutes that it took Weaver to return to Earth by his unplanned route, rumours of something unusual in the skies above had started to reach various ears – not all of them military. The first inkling of something awry came from radio chatter by airline pilots flying routes across eastern New Mexico at the time and who had seen the falling debris from the SR-71 but didn't know what it was.

'Albuquerque [Air Route Traffic Control] Center, this is American 85. Do you have anything on radar off to our right? Looks like some kind of a missile or something …'

'American 85, affirmative. I do have a radar-identified target.'

'Uh, what was it, some kind of missile?'

'American 85, I'm afraid I can't say.'

'What do you mean? It's a secret deal of some kind?'

'American 85, you're free to draw your own conclusions, sir.'

From another pilot: 'That's all right, American. TWA 12, we can't see it either.'

'Well, we could see it. It's real clear out here.'

'Well we can, too. I was just being facetious.'

At Edwards Air Force Base, they already knew that something bad had happened. All contact with DUTCH 54 had been lost – and that could mean only one thing.

On the ground, Weaver was struggling with his chute, which was billowing in the wind and proving hard to collapse with only one hand free, his other still occupied in holding up the troublesome faceplate. But this was a day for surprises and small miracles. From behind him a voice asked: 'Can I help you?' – not the first thing you expect to hear after having landed in the middle of nowhere. The voice belonged to ranch-owner Albert Mitchell Jr, who had been branding colts when he heard an explosion high above and shortly after saw two parachutes descend. Jumping into his two-seater helicopter, in a hanger nearby, he had flown out to the scene and found the first of the airmen. With Mitchell's help, the chute was soon collapsed and secured. Next Weaver freed himself from his parachute harness and found out what had been making the flapping noise on the way down: it was the shredded remains of his seatbelt and shoulder harness, still latched. At that moment, he realized the incredible truth – that his ejection seat had never left the plane and that he had simply been torn out of the wrecked cockpit by the extreme forces of the break-up.

Those forces had not been kind to Weaver's crew mate. Mitchell flew his helicopter out to check on him and came back a few minutes later with the grim news that Zwayer was dead, his neck broken. Mercifully, he must have died in an instant, even before he was thrown clear of the destruction.

A sixty-mile flight in Mitchell's helicopter brought Weaver to the hospital in Tucumcari, from where he contacted Lockheed's flight test office at Edwards Air Force Base to explain what had happened. The team had already heard about the Blackbird's loss and, knowing how fast and high it was travelling at the time, assumed there couldn't have been any survivors. US Air Force security and accident-response teams left immediately for the crash site to begin the long task of investigating the accident and collecting the debris that was scattered over an area fifteen miles long by ten miles wide.

A couple of weeks later, Weaver was back in an SR-71, flying the first mission of a brand-new plane at Lockheed's Palmdale assembly and test facility. Perhaps his fellow crew member behind him was a little edgy about the mental state of someone who had recently been through such a trauma, especially since the pilot isn't visible from the rear seat. Suddenly, just as the plane was about to take off, Weaver heard an anxious voice over the intercom.

'Bill! Bill! Are you there?'

It turned out that a red indicator saying 'Pilot Ejected' had lit up, owing to a faulty switch, on the master-warning panel in the rear cockpit just as the plane's front wheels had left the runway.

'Yeah, George. What's the matter?'

'Thank God! I thought you might have left.'

# 12

## FANTASTIC VOYAGE

Across the dark skies of central Africa flew a strange, slender craft and its two occupants, Dick Rutan and Jeana Yeager, squeezed uncomfortably into a space no larger than a telephone booth. Neither had slept well for days, and both felt ill. They were constantly concerned about the weather, their fuel supply, the state of their equipment, and keeping their unique vehicle on an even keel. But the remarkable goal of their mission kept them going: to make the first around-the-world flight by an aircraft without stopping or refuelling.

In 1949, a B-50 Superfortress, called *Lucky Lady II*, completed one whole trip around the planet without landing but was refuelled by air tankers several times en route. In 1962, a B-52 was flown for a then record distance of 12,532 miles without refuelling – well short of a circumnavigation. In the eyes of the Fédération Aéronautique Internationale, the minimum distance to qualify for a world flight had to be at least the distance around the Earth at the latitude of the Tropic of Cancer or Capricorn (about twenty-three degrees north or south of the equator), equal to 22,858 miles.

## PLANE CRAZY

One day in early 1981, at a table in a restaurant in Mojave, California, sat three people whose lives revolved around aviation. Burt Rutan was an aeronautical engineer with a reputation for innovative lightweight designs who headed a small aircraft company with his brother, Dick, a pilot with a long and distinguished career in the US Air Force. The other person present was Dick's wife, Jeana Yeager, another experienced pilot (unrelated, as it happens, to Chuck Yeager, the first human to break the sound barrier).

Burt had a proposition. How would Dick and Jeana feel about taking on one of the last great challenges in powered flight: to be the first to circumnavigate the globe in a plane without refuelling? Up to that point, no one had considered such a journey possible. A conventionally built aircraft loaded with enough fuel to fly 23,000 miles or so, non-stop, would weigh too much to get off the ground. But times had changed. New materials had become available that were just as strong as aluminium – the metal from which most airframes were made – but much lighter. Using such materials, known as composites, and an innovative design, Burt believed he could create a plane that was up to the task. On a napkin he sketched out his vision for the vessel, eventually to become known as Voyager.

Over the next five years or so, Voyager's design evolved and became reality. This was an aircraft like no other – built not for speed or for any reason other than staying in the air long enough to go once around the world. The plane itself, including airframe and engines, weighed a mere 2,250 pounds, about the weight of a compact car. Fully laden with fuel, it weighed about four times as much.

**34** Voyager circling before landing at Edwards Air Force Base.

The main wing was thin and very long: at 111 feet, it was longer than that of a 727 jetliner. In front of it was a smaller wing called a canard, which was a signature feature of many of Burt Rutan's aircraft designs. The central part of the plane held the crew, and on either side were two long outriggers serving as fuel tanks, so that the whole craft resembled a catamaran without the sail. With its big tanks and hollow wings, Voyager was, to a good approximation, a flying reservoir of gasoline. Seventeen separate containers were crammed into every possible space. In flight, the crew would regularly have to shift fuel from container to container to keep the plane balanced. Keeping the weight down was imperative: for every non-fuel pound added to the weight of the basic fuselage and wing, six more pounds of gasoline would be required. It was a good thing that both pilots were lightly built and that Jeana's presence added a mere ninety-eight pounds to the total.

One engine provided power, connected to a propeller at each end of the central body of the plane. At the front was a Continental O-240 four-cylinder air-cooled engine producing 130-horsepower, and at the back a slightly smaller IOL-200 putting out 110-horsepower. The former would be used only for take-off and gaining altitude, and would be turned off while cruising. Both engines had the advantage of burning fuel sparingly and being able to run reliably for long periods without maintenance – crucial given the length of the journey ahead and Voyager's paltry top speed of just 122 miles an hour.

## SHOESTRING AND COMPOSITES

Not since the Wright brothers had anyone attempted to break a major world record in a plane that they had made themselves from scratch. Dick Rutan and Jeana Yeager were the pilots, and they also volunteered to be the main builders, with Burt acting as chief designer.

But there was a problem: money. Dick and Jeana had $10,000 between them but with the high-tech composite materials and other pricey equipment that the aircraft needed the eventual price tag would be around $2 million. Sponsors were hard to come by. Many in the business world thought that the project was too risky, or simply not feasible. No one was willing to gamble on what seemed like little more than a wish and prayer. Government and big aerospace companies shied away from the venture completely. The brutal bottom line was that if Voyager came down early and ignominiously – especially if it ended up as a pile of wreckage with two dead pilots aboard – it wouldn't look good to have your logo showing on a piece of the debris.

So, Dick and Jeana just got to work with what they had, and what they could scavenge from friends, in a building owned by Burt at Mojave Airport. At first, progress was slow. But as word got around the close-knit aviation community of the south-west, curious folk started to stop by to see what was taking shape. Some visitors gave small donations to the cause; others were professionals who offered their expertise and, in some cases, eventually joined the Voyager team as advisers and manufacturing assistants. They included Bruce Evans, an expert in composites who became chief consultant, and Mike Melvill, a specialist in the same field who was also an employee of Burt Rutan's newly formed Scaled Composites LLC.

Voyager emerged from the drawing board in spring 1982. For the aircraft to have a structural weight of under 1,000 pounds, lightweight composites would have to be used throughout ninety-eight per cent of its wings and fuselage. Paper-thin sheets of carbon fibres were laid over a sandwiched layer of Nomex honeycomb – a resin-impregnated paper with similarities to bees' honeycombs – surrounded by two layers of carbon filter cloth containing epoxies. Graphite fibre, five times as strong as steel but weighing only half as much, gave strength, flexibility and lightness.

Weight was at an absolute premium. Burt Rutan insisted that Voyager should have no gadgetry other than what was essential to keeping the plane airborne. Dick said he was damned if he was going to cross oceans in an aircraft without radar, and forced his brother to concede on that one point. Otherwise Voyager was stripped to the bone. It lacked the normal safety equipment of modern planes. It didn't even have sound insulation to stop the incessant loud drone of the engines from penetrating the cabin. Doctors feared that, because of this, the pilots might suffer a permanent thirty per cent hearing loss. Fortunately, the Bose company came forward and provided two sets of special noise-reduction headsets for the flight.

There were no concessions to comfort either. The compartment in which Rutan and Yeager would be cooped up for more than a week measured just over 3 feet wide by 7.5 feet long. The person flying the plane – normally Rutan – sat in the pilot's seat. The other person had to lie down at all times. Even changing places was a gymnastic challenge and the toilet facilities can best be described as rudimentary. On the menu was nothing more exciting than food supplements, such as powdered milkshakes, and for exercise, which was more or less impossible anyway, there was a five-foot rubber band.

## AT THE MARGINS OF SAFETY

What really mattered was how Voyager would fly. Could it really stay aloft for 23,000 unbroken miles? Dick Rutan took it for its maiden flight, lasting forty minutes, on 22 June 1984, running with only a small fuel load. That experience pretty much confirmed what the team already knew: Voyager was frail, handled about as poorly as an aircraft can without falling from the sky, and was rough to ride. Big tails and rudders make for easy-to-fly planes and Voyager, in order to keep drag to a minimum, had neither of these. Everything on it was optimized for range. The control surfaces were small, flaps non-existent, and the structure pared back as much as possible to allow for more fuel.

On the second outing, Rutan ran into some thermals and got a taste of how insecure life aboard Voyager could be. The turbulence caused the long, thin wings to flex and bend, so that the plane alternately sank and climbed. Yeager, along for the flight this time, was reminded of being on a sailing boat, rising and falling on a lively sea, and for the first time in her life suffered from motion sickness.

The ends of the slender main wings were designed to be able to move up or down as much as thirty feet and could take a lot of stress. But still the team feared that they might flex too much and simply snap off, sending the plane to its doom. On a later flight, Rutan was convinced that they had come within seconds of disaster. The pitching was so bad, and the wing oscillations so extreme, that it took all his skill to escape the thermal and bring the plane in for a safe landing. One thing was clear: Voyager would have to avoid as much rough weather as possible on its round-the-world flight.

Dick Rutan admired the beauty of the plane – it was one of the most aesthetically pleasing vessels ever to take to the air. He was also aware that it was inherently unsafe and that he and Yeager risked their lives every time they went up in it. In fact, throughout the preparations for the great adventure, Rutan had a premonition that Voyager would crash and he would be killed. If the world trip was successful, he firmly believed that its next and final journey should be to the Smithsonian to be put on display to the public.

After numerous test flights, some hairier than others, Voyager was finally set to go. Normally the best time, in terms of the weather, for attempting a round-the-world flight is June to August. But the team was impatient to be off without further delay – a brave decision given how poorly Voyager performed in anything but the calmest air. On 14 December 1986, Rutan and Yeager walked around the plane for one final check before boarding – Rutan into the only seat and Yeager onto the floor alongside him. They were ready to go.

## NINE DAYS IN THE TORTURE CHAMBER

Flight controllers at Edwards Air Force Base, from where the insect-like plane would start its mission and to where, hopefully, more than a week later, it would return, gave the all-clear for take-off. Voyager had never been so heavy before – fully laden with fuel, provisions, equipment and crew. The spindly wings were so weighed down with gasoline that their tips almost touched the ground. It was just after 8 a.m. when Rutan released the brake and eased the throttles forward. Slowly the aircraft began to move, gathering speed, but not quickly enough. The ends of the wings weren't lifting; in fact, they were now scraping the tarmac, damaging their ends. Without sufficient lift, Voyager was approaching the end of the runway and seemed as if it might crash. 'Pull up on the stick!' yelled Burt Rutan over the radio. But Dick didn't hear the warning, nor did he know about the problem with the wings. Looking straight ahead he was focused simply on keeping the plane straight and picking up enough speed to get airborne. At the last moment, the long wings swept up, bearing Voyager into the air at the start of its marathon journey. But the plane and its pilots had already broken one record: having used up all but 1,000 feet of the longest runway in the world, they had just made the lengthiest take-off ever from Edwards Air Force Base.

# RUTAN VOYAGER

**Crew:** 2
**Length:** 29 ft 2 in (8.9 m)
**Wingspan:** 110 ft 8 in (33.8 m)
**Empty weight:** 2,250 lbs (1,020 kgs)
**Maximum speed:** 122 mph (196 km/h)
**Range:** 24,986 miles (42,212 km)

Air-cooled Teledyne Continental O-240 engine (130 horsepower)

Habitable section of the aircraft had one seat and a makeshift bunk

Water-cooled Teledyne Continental IOL-200 engine (110 horsepower)

**35** Voyager specifications.

Dick Rutan learned from the chase plane – a twin-engined Beech Duchess carrying his brother Burt and two others – of the extent of the damage to his wings, and angled the plane slightly so that the force of rushing air would clean up the broken ends. Then he aimed Voyager out over the Pacific Ocean.

Meanwhile, Jeana Yeager had her work cut out raising the landing gear. There was no hydraulic system to do this automatically: that was a luxury the weight-obsessed design couldn't afford. Instead, she had to laboriously crank a primitive sail boat winch for fifteen minutes, slowly lifting the three wheels into position one at a time.

Living and working within the claustrophobic interior of Voyager was hellish; the two pilots had nicknamed the space 'the torture chamber'. At the best of times, being cooped up and uncomfortable, squashed together so close, would have caused tension between two people. But in Rutan and Yeager's case it was made worse because the couple had separated only a few months earlier. Their marriage had broken down over the very thing that had drawn them together: their passion for flight. Each was so intensely motivated that they competed for control of the aircraft they flew together, and that competition to be top dog had spread to their relationship. Given the other pressures on them, the last thing they needed was bad weather. But, inevitably, on such a long trek, it came – and on only the second day.

They were heading directly for tropical storm Marge. The powerful winds packed by Marge could easily make matchwood of Voyager's feeble airframe. Fortunately, weather advice feeding through from the mission's weather expert was accurate. The storm was a known obstacle and the original plan was to fly between two of its pinwheel arms, in a region of relatively still air. But then Marge veered north faster than expected, towards the Philippines, so that on its present heading Voyager would run straight into it. To add to the excitement, a low-pressure system was moving in from the north. Between this aerial Scylla and Charybdis was a narrow passage of calm air towards which Voyager was directed to head in the hope that it would make it through before the channel closed. With no other option – the weather systems were too wide to fly around – Voyager drove ahead, a bank of stormy weather to port and towering clouds to starboard, hugging the left wall of Marge. It worked. Subjected to some rain but little turbulence, one of the least airworthy flying machines of the late twentieth century came through unscathed and even managed to pick up a healthy tail wind, which sped it along between the islands of Guam and Saipan.

Over Africa, more stormy weather struck, adding to the pressure on Rutan, who had been at the controls for almost two and half days straight and was dog tired. For brief periods he would swap places with Yeager, with the plane on autopilot, just to catch a half-hour's rest here and there. Fatigue was starting to lead to mistakes. A red warning light signalled that one of the engines was short on oil. The pilots had been so busy making sure that they avoided the worst of the turbulence, not to mention occasional high mountains while crossing Africa, that they had forgotten to keep an eye on the oil levels. Luckily, they hadn't left it too late and were able to add the extra lubricant before the engine was damaged.

A fuel leak, eventually traced to some new valves that had been installed, was another cause of concern. Even with the loss, a computer confirmed that there ought to be enough fuel aboard to make it home. But there was one final moment of doubt. Flying up the coast of Mexico, on the last leg of the long journey, the rear engine failed. With the more powerful front engine already shut down to save gas, the aircraft quickly began to lose speed and altitude. For five minutes Voyager plunged. Finally Rutan managed to spark the front engine into life and then the rear one.

As Voyager came along the California coast towards home, Burt was up in the chase plane with Mike Melvill, anxiously looking for any sign of the long-distance traveller. Then they spotted what they thought was the slender aircraft in the pre-dawn light, illuminated by its flashing strobe.

'Dick, turn off the strobe light,' Burt radioed his brother. The light went out, and both Rutan and Melvill broke down in tears at the sight of their friends coming home.

Nine days, three minutes, and forty-four seconds after struggling to take to the air from Edwards Air Force Base, Voyager returned to a smooth landing. It had flown 26,366 miles, all the way around the Earth, without stopping and without taking on fuel. Dick Rutan, after stepping out and stretching his legs, said: 'This was the last major event of atmospheric flight.'

# 13

## JETMAN

Even James Bond would have blenched at the prospect. But to Swiss inventor and aviator extraordinaire Yves Rossy, it was just another day at the office. On 16 September 2008, Rossy, known variously as Jetman, Rocketman, and Fusionman, stepped out of a light plane more than 7,500 feet above Calais, France, ignited four miniature jet engines attached to the carbon-composite wing on his back, and hurtled out across the English Channel at speeds of up 186 miles per hour. He completed the twenty-two-mile journey to Dover in a shade over nine minutes before landing via parachute. Rossy's jet-pack is the latest and most extreme evolution of what is known as wingsuit flying – a modern-day version of attempts to fly like a bird.

### THE FIRST BIRDMEN

For centuries, early pioneers of flight tried to take to the air by fixing artificial wings to their arms and backs. But just as the myth of Icarus ended with the Greek hero tumbling to his death, history books are filled with tales of would-be birdmen

**36** Yves Rossy, aka 'Jetman' or 'Rocketman', flying with his jet-propelled wing.

leaping from high places, wings spread wide – and falling disastrously back to Earth.

The first birdmen to achieve any kind of consistent success emerged in the 1930s as skydiving became increasingly popular. Especially well known was the American airshow daredevil Clem Sohn. His wings were made from tough aircraft canvas, mounted on a steel framework, which formed a web between each arm and the side of his body. The wings were designed so that they couldn't open too far and tear his arms out of their sockets. Another sheet of cloth between his legs acted like a bird's tail. The shape of his wings and the large goggles he wore inspired his nickname 'The Batman'.

Jumping from a plane at altitudes of up to 18,000 feet, Sohn would glide down, swooping, banking and looping, before opening his parachute only 800 or 900 feet above the ground. He made a good living doing his spectacular stunts at fairs and air meets, but it was only a matter of time before something went badly wrong. During a jump at the opening ceremony of London's Gatwick Airport on 6 June 1936, his main parachute got entangled in his wings. He was less than 200 feet up when he finally managed to open his emergency parachute, and, although he landed heavily, he escaped with a broken shoulder. On 25 April 1937, at a show in Vincennes, France, his luck ran out. Before taking off, he had said, 'I feel as safe as you would in your grandmother's kitchen.' But during the descent his main parachute failed to open. A crowd of 100,000 people watched him frantically tug on the ripcord of his emergency chute, but that also failed, and Sohn, only twenty-six years old, plunged to his death.

The fatality rate of the early birdmen doesn't bear thinking about. Between 1930 and 1961, of seventy-five individuals who tried to push back the frontiers of human winged flight, seventy-two died in accidents. Meanwhile, technology progressed and the lure of soaring like a bird remained as strong as ever. A 1969 film called *The Gypsy Moths*, starring Burt Lancaster, showcased the exploits of barnstorming skydivers and the sport of what became known as wingsuit flying. Although the movie wasn't a huge commercial success, it helped inspire a new generation of birdmen who would finally achieve the age-old dream to fly like a bird.

## WINGSUIT FLYING COMES OF AGE

There isn't much to a wingsuit – mainly just fabric stretched from the arms to the body and between the legs – but getting it just right, and making it of the right stuff, is key. The basic idea is to increase the surface area of the human body to achieve more lift, so you end up looking, and hopefully gliding, something like a flying squirrel. The only snag is that you can't land with a wingsuit (yet) because it can't slow down the vertical descent enough, unless you are travelling very fast horizontally, so a parachute is essential for making a safe landing. A lot of care and attention has to go into designing the wingsuit-parachute combination so that when the chute finally does deploy, it opens cleanly and doesn't get snagged in the panels of the suit.

## PATRICK DE GAYARDON

At the cutting edge of the modern movement of skydiving in the 1990s was Frenchman Patrick de Gayardon. He used a snowboard, attached to his feet, to take skydiving to a new level known as skysurfing, and also came up with a revolutionary wingsuit design. The wings of de Gayardon's suit, which inflated in the same way as canopies used in all kinds of contemporary skydiving and paragliding, had a bottom and top surface, linked by ribs. The suit incorporated a safety system, allowing the wings to separate via a wire running down the wing and body.

De Gayardon showed the extraordinary manoeuvrability and control he could get from his bat-like suit during some incredible jumps off famous locations such as Mont Blanc and the Grand Canyon. He was also the first person to fly in a wingsuit close to the ground when he skimmed just yards above the famous terraces of the Aiguille du Midi, under the noses of journalists. In 1997 he jumped out of a Pilatus PC-6 Porter light aircraft, glided down, with the plane diving alongside him, both descending at about forty-five degrees and slowly edging close together, before finally de Gayardon flew back through the open door of the cabin. Sadly, the Frenchman died in 1998 during a sunset skydive in Hawaii while testing a modification to his parachute container, the accident being attributed to a rigging error.

**37** Aiguille du Midi ('Needle of the South') in the French Alps over which Patrick de Gayardon flew in 1997.

Not until the mid-1990s did wingsuits step out of the era of makeshift materials and pure daredevilry to become a science in themselves. Wingsuit flying came of age and many individuals and groups entered the field, trying to improve the methods and tools of the sport and push back the envelope of what was possible. In 1999, Croatian jumpsuit designer Robert Pecnik, together with Jari Kuosma from Finland, formed Birdman – the first company to make wingsuits commercially available.

In 2005 a new subgenre of the sport was born when a group of Norwegian BASE jumpers released video footage of its members wingsuit flying almost within touching distance of cliff faces.* Very quickly, so-called 'proximity flying' became one of the most popular activities within BASE jumping. Other enthusiasts used wingsuits to set new time and distance records. Again in 2005, three Spanish skydivers jumped from 35,000 feet and crossed the Strait of Gibraltar in wingsuits, covering 12.5 miles and setting a record free-fall time of 6.05 minutes. But even this achievement was soon to be eclipsed.

## THE JET-PROPELLED MAN

Yves Rossy had gained plenty of experience of flying at high speed even before he tucked a bunch of jet engines under his arms and flicked the switch to turn them on. He was a fighter pilot for seventeen years in the Swiss Air Force, spending his days behind the controls of Mirage IIIs, Northrop F-5 Tiger

---

* 'BASE' stands for 'buildings, antennas, spans (bridges), and earth (rock faces)'.

IIs, and Hawker Hunters, before going on to fly DC-9s and Boeing 747s for Swissair and, later, Swiss International Air Lines. But deep inside Rossy, there was always a birdman trying to get out.

Extreme sports were in Rossy's blood – everything from barefoot water-skiing, wake-boarding, and hydro-speeding, to hang-gliding, paragliding, and flying acrobatics. He started skydiving in 1990 before moving on to sky-surfing and earning himself a place in the *Guinness Book of Records* with a sky-surf from a hot-air balloon over the Matterhorn. The years 1997 to 1998 saw him building and trial-testing sky-surf wings with

**38** Yves Rossy.

the help of researchers from the Ecole d'Ingénieurs de Genève. Then he began work on the technology that would ultimately bring him world fame.

Rossy wanted to go down a different route from fabric wingsuits. Maybe it came from his many years of piloting so many different kinds of aircraft. But he wanted to fly with wings on his back – effectively turning himself into a living plane. So he joined forces with two high-tech companies, Prospective Concepts and ACT Composites, to make a pair of wings that would allow him to leap from a light aircraft at several thousand feet and then fly like a glider until, when close to the ground, he would deploy a parachute to land. For several years he and these two firms worked on perfecting wings that were part inflatable and part rigid. The so-called 'Flying Man' project culminated, in 2002, in Rossy successfully crossing Lake Geneva from Evian to Lausanne with his aerial apparatus.

But that was just an aperitif as far as the Swiss pilot was concerned. The next step was to motorize his wings to be able to fly straight and level for miles, or even climb, like Superman or the Rocketeer. What he needed were engines light enough for a person to be able to carry but powerful enough to give a good boost for ten minutes or more. In 2002, Rossy got in touch with the German-based company JetCat, known to enthusiasts as one of the world's leading manufacturers of jet engines for model aircraft. It was a perfect match-up and, in the end, JetCat's kerosene-fuelled P200 engines proved ideal for the task. A lot of groundwork had to be done, testing different engines and air intakes, and then fine-tuning the chosen P200s so that they would perform well many thousands of feet up – at altitudes way beyond anything they had been flown at before. A special electronic starter system was devised to ensure that all the engines ignited in the same split second, otherwise Rossy might be thrown into an uncontrollable spin. Aligning the engines precisely when they were mounted to the wing was also crucial to avoid any instability.

The wings themselves had to be adapted to cope with the new demands of carrying jets. They were lengthened and made more rigid by having their inflatable component removed. Handles were fitted to them so that Rossy could electronically manipulate the wing tips, giving him the freedom to decide when he wanted to glide or dive.

Finally, the preparations were over. At 7.30 p.m. on 24 June 2004, Rossy stepped out of a Pilatus plane, some 12,000 feet above Yverdon airfield, near Vaud, Switzerland. After a few seconds he pulled on the small lever that controlled the opening of his wings, and glided down to 7,500 feet before igniting his engines. Once he was sure they were running smoothly, he throttled them up, pushing his speed to 115 miles per hour. For four minutes he flew horizontally, about 5,000 feet from the ground, realizing another dream of human flight – not only to glide and dive like a bird but to fly level and have the power to soar up into the sky. Rossy even remembered to turn on his smoke generators, leaving behind a temporary trail of his successful venture. 'It was absolutely fantastic,' he said after the flight. 'Freedom in three dimensions ...'

Further improvements came to the design of the jet-wing, leading to a configuration similar to that used by Rossy today, with a span, when open, of just over seven feet, and a weight of 120 pounds, including four P200 engines. Firing all at once, on full throttle, the jets supply around 200 pounds of thrust for ten minutes until the fuel runs out. An insulated suit, in combination with the chilly air in which Rossy usually flies, is enough to protect him against the heat produced by the engines. To land, he opens his main parachute, folds his wing, and allows it to fall supported by its own rescue chute.

## ACROSS CHANNELS AND CANYONS

Images of the Swiss pilot, coolly looking sideways at the camera while hurtling across the sky, soon adorned newspapers and magazines around the world. In 2008, the same year that Rossy followed Louis Blériot's route across the English Channel, he flew high above the Alps peaking out at 189 miles per hour.

The following year, the modern-day Icarus sought to cross the Strait of Gibraltar, in the hope of becoming the first person to fly between two continents using a jet-pack. It all started well with a jump from a light plane about 6,500 feet above Tangier in Morocco, heading in the direction of Atlanterra in Spain. But part-way across, Rossy met with strong winds and banks of cloud and was forced to ditch into the sea, three miles from the Spanish coast. Fortunately, his support helicopter rescued him ten minutes later and flew him to a nearby hospital, from where he was later released in good health. His jet-pack was unfortunately not recovered by the Spanish coast guard, despite its float having deployed after it touched down in the water.

In the last couple of years, Rossy's stunts have become even more extreme. In November 2010 he dropped from a hot-air balloon at 7,900 feet, performed two vertical loops, and stayed aloft for a total of eighteen minutes. Six months later he flew across the Grand Canyon in Arizona, after the

**39** Rossy flying over the Grand Canyon.

US Federal Aviation Administration agreed to classify his jet-wing system as an aircraft, and waived the normal flight-testing time of twenty-five to forty hours. After soaring 200 feet above the canyon rim and reaching speeds of up to 190 miles per hour, he opened his parachute and landed safely on the canyon floor.

What could Rossy do next to top all these achievements and fantastic stunts? The answer came on 26 November 2011, when he flew in formation with two Aero Vodochody L-39C jets from the Breitling demonstration team. Granted, he was pushing his powered wing to the limit, while the L-39Cs were only just above their stall speed, but it was a jaw-dropping sight – and testimony to the extraordinary control, using only

**40** Rossy flying in formation with two jets from the Breitling demonstration team.

subtle movements of his body, that Rossy now has over his aerial apparatus. In fact, the system is so responsive and reactive in flight that any movements of the pilot's head, arms and legs have to be small and carefully done in order to avoid going out of control.

The military was so impressed by the powered wings that it asked for prototypes, but Rossy has insisted that his invention is for use only in sport and recreation. But with a cost to build of around $250,000, it is unlikely that the jet-wing pack will fall within the budget of many enthusiasts.

# 14

## FALLING HERO

The tall, tubby guy with the somewhat gaunt face and handlebar moustache riding the elevator to the eighty-sixth floor may have looked a little odd. But you see all sorts in a big city. What he did next, though, was distinctly out of the norm – even for New York. Having made his way to the observation deck of the Empire State Building, he shed his fake facial hair and outer garments, revealing a parachute inside his fat suit. Then he scrambled over the safety fence, intent on leaping over the side into 1,050 feet of thin air. At the last moment, a couple of security guards grabbed hold of him through the rails and he was arrested and charged with 'reckless endangerment with depraved indifference to life'. The last part of the charge he considered a bit unfair. After all, he had meant to time his jump meticulously with an eye to landing in Fifth Avenue when the traffic lights were on red.

It was 27 April 2006, and the then 29-year-old Jeb Corliss wasn't exactly an unknown even before his detainment on the Big Apple's most famous edifice. In BASE-jumping circles his exploits were already legendary, and he was known to a wider audience through his hosting of the Discovery Channel series *Stunt Junkies*. Discovery promptly dispensed with his services following his arrest by the NYPD, but, publicity-wise, the aborted rapid descent from the Empire State Building could hardly have gone better for the young daredevil. Every TV and radio chat show wanted him to retell his story, and his picture made the front page of the *New York Times*.

Sometimes the best publicity comes when bad things happen to you. Corliss – never one to shrink from sprinkling a few 'I's in his conversation – basked in his depiction as an anti-hero. In fact, dressed in his all-black wingsuit, and wearing his usual severe expression, he exudes the appearance of a dark superhero from some American comic-book series. His sentence – three years' probation plus one hundred hours' community service – was overturned by a Manhattan state judge on the grounds that Corliss 'was experienced and careful enough to jump off a building without endangering his own life or anyone else's' – before being affirmed early in 2009. It may be that he was secretly pleased when he was later banned from ever setting foot in the world's most famous skyscraper again.

## BATMAN BEGINS

Corliss had an upbringing that almost guaranteed he was going to be offbeat in some way. His parents dealt in exotic artefacts and roamed the world with their three kids in tow, looking for obscure items to sell in their Santa Fe gallery. The young Jeb was pretty much left to his own devices. As a five-year-old, living temporarily in Nepal, he would drift over to the local Buddhist monastery and hang out with the monks.

His parents divorced when he was eight and Corliss, now living with his mother and sisters, was pulled out of school before he hit seventh grade. He came to consider school to be evil, a place fit only for fighting. As a young teenager, he was angry, disaffected, depressed. It could all have gone horribly wrong for him but, fortunately, he discovered skydiving – an outlet for his frustrations. But even then, as an eighteen-year-old, he knew he wanted to do more than just skydive. He wanted the extra adrenaline rush that came from BASE jumping, a relatively new sport in 1994. Such was his dark mental outlook

**41** BASE jumping from an antenna.

at the time that nothing held any fear for him, and after just a few weeks of training he wanted to jump from an antenna – the 'A' of BASE – one of the toughest choices he could have made. Not his mentors in the sport, nor even his mother, could dissuade him from going ahead.

For his jump, Corliss picked a 300-foot tower near Camarillo in Ventura County, California. It was a ridiculously low height for a beginner, barely giving even an expert the chance to open his chute in time to make a safe landing. On his own, at night, the troubled teen drove out to the tower, climbed over the razor-wire fence surrounding it, and began to ascend the tower's ladder. All the time, as he went up, he told himself that he wouldn't go through with it – that he would look around when he got to the top and then take the slow way back down.

Except it didn't work out that way. Corliss found himself leaping and falling, and at the same time tugging to release his pilot chute. Nothing happened. Already he was flashing past the red light that marked the half-way point of the tower, with the ground racing towards him. Finally his parachute opened with a bang and, seconds later, he hit the road, ending up on his back, winded, but miraculously unhurt.

## DEADLY BUSINESS

An experience that would have put most people off BASE jumping for life – assuming they were crazy enough to try it in the first place – merely encouraged Corliss. Five years later, and a lot more savvy, he persuaded the producers of a show called *Real TV* to film him diving into the 310-foot-high Howick

Falls, in KwaZulu-Natal province, South Africa. Part-way down he veered off course, exposing his chute to the tumbling torrent, and causing it to collapse and him to plummet into the rocks below. He survived, barely, but suffered several cracked vertebrae and the attentions of various aggressive animals in the water at the base of the falls. A month in hospital did

**42** Steph Davis performing a BASE jump in a wingsuit.

nothing to dampen his spirits. On the contrary, the footage of his spectacular jump and near-fatal crash made for compelling viewing, and he saw a new lucrative business opportunity for himself and other BASE jumpers – selling the rights to videos of his stunts to documentary producers and extreme-sports shows around the world.

Corliss gained an international reputation for his agility and acrobatic skill as a wingsuit flier. His customized wingsuit has lightweight, ripstop nylon panels from the undersides of his arms to his torso and between his legs. Air inlets on the wing's leading edge allow it to inflate and stay rigidly pressurized for flight. Glide ratios of three feet forward for every one foot down are possible, but Corliss often flies 1:1, descending at a forty-five-degree angle, which gives him the option of pulling up sharply and changing his angle of approach.

But whatever an individual's level of expertise, tragedy is never far away in a sport where those involved rely on split-second timing to avoid disaster. In 2003, Corliss and a close friend, popular Australian BASE jumper Dwain Weston – one of the best in his sport – attempted a combo jump from a plane flying above Colorado's 1,053-foot-high Royal Gorge Bridge, the world's highest suspension bridge. The original idea was for Corliss to wingsuit and Weston to parachute, but at the last minute Weston decided to wingsuit as well – Corliss flying under the bridge and Weston going over it, in opposite directions.

Weston swooped in first, aiming to miss the railing by a few yards, with a couple of hundred spectators looking on. But the gorge is steep and narrow, and the winds around it are tricky. The thirty-year-old Australian misjudged his distance and height by a fraction and smacked into the railing at around one hundred miles per hour, tearing him in two. As Corliss approached from the other side, his first thought was that people were throwing things off the bridge, but then he realized in horror what the falling pieces were. When he landed he was covered in his friend's blood.

In the close-knit community of BASE jumpers, extraordinary exhilaration and extreme trauma go hand in hand. Many participants live life for the moment and accept that, even with the best preparations and equipment, they may quite easily die on their next leap. Corliss has said openly that he expects to be killed eventually if he carries on. Yet, perhaps not surprisingly for someone who has been diagnosed with counterphobia – a pathological desire to confront fear – he has continued to make spectacular leaps off some of the world's most iconic structures, including the Golden Gate Bridge, the Petronas Towers in Malaysia, Seattle's Space Needle, and the Stratosphere casino in Las Vegas, having, in the last case, avoided the suspicions of security guards by hiding his parachute inside a large soft toy. In 2011, he performed his most breathtaking stunt to date, jumping out of a helicopter at 6,000 feet, then hurtling through a 100-foot-wide natural archway in Tianmen Mountain, Hunan Province, China, before landing by parachute on a nearby bridge.

## THE GREATEST CHALLENGE

In 2009, a BASE jumper wearing a wingsuit hit the side of a mountain in Russia but then inadvertently almost did what no other jumper or wingsuiter has ever done: he had left it too late to open his parachute, so struck the mountainside almost at terminal speed – the highest speed reached during free fall. However, the slope was so steep and covered in snow, before levelling, that the jumper survived the long tumble – albeit with many broken bones. It was the closest any skydiver or wingsuit flier had come to slowing down gradually from terminal speed and landing safely without the help of a parachute.

True, there had been a variety of devices, called jet-packs or rocket-belts, like the one made famous in the James Bond film *Thunderball*, which enabled vertical take-offs and landings. But these didn't have any lifting surfaces to allow gliding or other kinds of movement made possible by wings. The challenge was to pull off a successful, safe landing wearing only a wingsuit.

Corliss aimed to be the first. But the scheme he devised wouldn't be cheap. What he needed were some sponsors to stump up about $3 million to build a temporary 2,000-foot-long landing ramp, which would run from the top of a structure in that glitziest of cities, Las Vegas, to street level. He envisaged the ramp being made from some kind of tensile fabric whose surface would be kept taut by a system of cables.

Corliss intended to drop from a helicopter hovering several thousand feet above the famous Strip. He would be aiming for a disturbingly small approach window at the top of the landing slope, just twenty feet square. Getting it spot on would be critical, because if he hit the ramp in the wrong place or at the wrong angle he might as well smash into a stone wall or belly-flop into the sea off a high cliff.

If all went to plan, Corliss hoped to touch down like a ski jumper, matching the angle of the slope with the angle of his body. The only differences would be that he would be making contact at one hundred miles per hour – about forty miles per hour faster than a typical ski jumper – and he would land not

**43** A wingsuit flier in Holland.

on his feet but on his rib cage. The ramp's fabric would be able to absorb some of the impact and friction, but it couldn't be too elastic and forgiving or Corliss would simply bounce off. To survive unharmed, Corliss would have to touch down on the forty-five-degree landing ramp with a flight angle no steeper than about fifty degrees.

Finding sponsors proved tougher than Corliss expected. Companies seemed less than keen to finance a venture that stood a fair chance of ending in disaster. In the end, though, Corliss's dream was ended not by lack of support but by a little-known British stuntman, Gary Connery, who, on 23

May 2012, became the first person to survive an intentional parachute-less jump. Leaping from a helicopter 2,400 feet above the Oxfordshire countryside, the wingsuited Connery swooped to a safe touchdown on a distinctively low-tech landing strip – a 350-feet-long by 45-feet-wide pile of cardboard boxes.

A few months before this, Corliss got a severe reminder of the perils of his occupation – as if any more were needed – following a BASE jump off Table Mountain, outside Cape Town, South Africa. Some 200 feet from the bottom, a gust of wind – or perhaps a slight misjudgement – brought Corliss too close to the rocks at the base of the mountain and he struck feet first. He broke both ankles, three toes, and a fibula, ripped the anterior cruciate ligament in his left knee, and opened up an ugly gash in front of his right shin, before spinning out and deploying his chute for a rough and excruciatingly painful landing. Having been airlifted out by a Red Cross Air Mercy service, he sent a message from his hospital bed: 'I feel better than I've ever felt'.

Little over six months later, Corliss was back in the air, training for new challenges. Of his long-time survival prospects, he's not overly optimistic. A recent post on his Facebook page reads: 'My death shall be violent, brutal and there will be blood …'

# FURTHER READING

Abrams, Michael. *Birdmen, Batmen, and Skyflyers: Wingsuits and the Pioneers Who Flew in Them, Fell in Them, and Perfected Them.* New York: Three Rivers Press, 2007.

Branson, Richard. *Reach for the Skies: Ballooning, Birdmen and Blasting into Space.* London: Virgin Books, 2011.

Brown, Eric. *Wings of the Weird and Wonderful.* Crowborough, East Sussex: Hikoki Publications, 2010.

Brown, Peter and Broeskc, Pat H. *Howard Hughes – The Untold Story.* New York: Dutton, 2004.

Burgess, Colin. *Selecting the Mercury Seven: The Search for America's First Astronauts.* Berlin: Springer-Praxis, 2011.

Caidin, Martin. *Barnstorming.* New York: Bantam Books, 1991.

Chambers, M.J. and Chambers, Randall. *Getting Off the Planet: Training Astronauts.* Burlington, Ontario: Apogee Books, 2005.

Cleveland, Carl M. *'Upside-Down' Pangborn: King of the Barnstormers.* Glendale, Cal.: Aviation Book Company, 1978.

Crouch, Tom D. *Wings: A History of Aviation from Kites to the Space Age.* New York: W.W. Norton & Co., 2004.

Dedijer, Jevto. *Base 66: A Story of Fear, Fun, and Freefall.* Bloomington, Ind.: iUniverse, 2004.

Dee, Richard. *The Man Who Discovered Flight: George Cayley and the First Airplane*. Toronto: McClelland & Stewart, 2007.

Dwiggins, Don. *The Barnstormers: Flying Daredevils of the Roaring Twenties*. New York: Grosset and Dunlap, 1968.

Fossett, Steve. *Chasing the Wind: The Autobiography of Steve Fossett*. London: Virgin Books, 2006.

Goodrum, Alistair. *Balloons, Blériots and Barnstormers: 200 Years of Flying for Fun*. Stroud, Glos., UK: The History Press, 2009.

Graham, Richard H. *SR-71 Blackbird: Stories, Tales and Legends*. Minneapolis, Minn.: Motorbooks International, 2002.

Grant, Reg. *Flight*. London: Dorling Kindersley, 2010.

Gwynn-Jones, Terry. *Farther and Faster: Aviation's Adventuring Years, 1909–1939*. Washington, DC: Smithsonian Institution Press, 1991.

Harding, John. *Flying's Strangest Moments: Extraordinary But True Stories from Over 1000 Years of Aviation History*. London: Robson Books, 2006.

Jenkins, Dennis R. and Landis, Tony R. *Hypersonic: The Story of the North American X-15*. North Branch, Minn.: Specialty Press, 2008.

Kittinger, Joe W. and Ryan, Craig. *Come Up and Get Me: An Autobiography of Colonel Joseph Kittinger*. Albuquerque, NM: University of New Mexico Press, 2010.

Mitchell, Charles R. and House, Kirk W. *Flying High: Pioneer Women in American Aviation*. Stroud, Glos.: The History Press, 2002.

Nevin, David. *The Pathfinders*. Alexandria, Va.: Time-Life Books, 1980.

Franks, Norman L.R. *Sopwith Triplane Aces of World War I*. Oxford: Osprey Publishing, 2004.

O'Neil, Paul. *Barnstormers and Speed Kings*. Alexandria, Va.: Time-Life Books, 1981.

Pisano, Dominick; van der Linden, Robert, and Winter, Frank H. *Chuck Yeager and the Bell X-1: Breaking the Sound Barrier*. Harry N. Abrams, Inc. 2006.

Rich, Ben R. *Skunk Works*. London: Sphere, 1995.

Roseberry, Cecil R. *The Challenging Skies: The Colorful Story of Aviation's Most Exciting Years, 1919–1939*. Garden City, New York: Doubleday, 1966.

Ryan, Craig. *The Pre-Astronauts: Manned Ballooning on the Threshold of Space*. Annapolis, MD: Naval Institute Press, 2003.

Schiff, Barry. *Dream Aircraft: The Most Fascinating Airplanes I've Ever Flown*. Newcastle, WA: Aviation Supplies & Academics, 2007.

Thompson, Milton O. *At the Edge of Space: The X-15 Flight Program*. Washington, DC: Smithsonian Books, 2003.

Wolfe, Tom. *The Right Stuff*. New York: Vintage, 2005.

Yeager, Chuck. *Yeager*. New York: Bantam, 1990.

# INDEX